...d Scope

..."Springer Theses" brings together a selection of the very best Ph.D. ...m around the world and across the physical sciences. Nominated and ...y two recognized specialists, each published volume has been selected ...entific excellence and the high impact of its contents for the pertinent ...earch. For greater accessibility to non-specialists, the published versions ...extended introduction, as well as a foreword by the student's supervisor ...the special relevance of the work for the field. As a whole, the series will ...aluable resource both for newcomers to the research fields described, and ...cientists seeking detailed background information on special questions. ...provides an accredited documentation of the valuable contributions made ...younger generation of scientists.

...re accepted into the series by invited nomination only ...t fulfill all of the following criteria

...ist be written in good English.
...c should fall within the confines of Chemistry, Physics, Earth Sciences, ...ring and related interdisciplinary fields such as Materials, Nanoscience, ...l Engineering, Complex Systems and Biophysics.
...k reported in the thesis must represent a significant scientific advance.
...esis includes previously published material, permission to reproduce ...t be gained from the respective copyright holder.
...ist have been examined and passed during the 12 months prior to ...on.
...esis should include a foreword by the supervisor outlining the signifi- ...its content.
...ses should have a clearly defined structure including an introduction ...e to scientists not expert in that particular field.

...mation about this series at http://www.springer.com/series/8790

Springer Theses

Recognizing Outstanding Ph.D. Re

Aim:

The s
theses
endors
for its
field o
includ
explair
provid
for oth
Finally
by tod

These
and n

- They
- The t
  Engi
  Chen
- The v
- If the
  this n
- They
  nomir
- Each
  cance
- The t
  access

More inf

David James Martin

Investigation into High Efficiency Visible Light Photocatalysts for Water Reduction and Oxidation

Doctoral Thesis accepted by
the University College London, UK

 Springer

Author
Dr. David James Martin
Department of Chemical Engineering
University College London
London
UK

Supervisor
Prof. Junwang Tang
Department of Chemical Engineering
University College London
London
UK

ISSN 2190-5053   ISSN 2190-5061   (electronic)
Springer Theses
ISBN 978-3-319-18487-6   ISBN 978-3-319-18488-3   (eBook)
DOI 10.1007/978-3-319-18488-3

Library of Congress Control Number: 2015939152

Springer Cham Heidelberg New York Dordrecht London

Printed on acid-free paper

Springer International Publishing AG Switzerland is part of Springer Science+Business Media
(www.springer.com)

Parts of this thesis have been published in the following journal articles:

1. **Efficient visible driven photocatalyst, silver phosphate: performance, understanding and perspective.** David James Martin, Liu Guigao, Jinhua Ye and Junwang Tang. *Chemical Society Reviews*, in review, **2015**
2. **Visible Light-Driven Pure Water Splitting by a Nature-Inspired Organic Semiconductor-Based System.** David James Martin, Philip James Thomas Reardon, Savio J.A. Moniz, and Junwang Tang. *Journal of the American Chemical Society*, **2014**, 136 (36), 12568–12571
3. **Highly Efficient H$_2$ Evolution from Water under visible light by Structure-Controlled Graphitic Carbon Nitride.** David James Martin, Kaipei Qiu, Stephen Andrew Shevlin, Albertus Denny Handoko, Xiaowei Chen, Zheng Xiao Guo & Junwang Tang. *Angewandte Chemie International Edition*, **2014**, 53 (35), 9240–9245
4. **Facet engineered Ag$_3$PO$_4$ for efficient water photooxidation.** David James Martin, Naoto Umezawa, Xiaowei Chen, Jinhua Ye and Junwang Tang. *Energy & Environmental Science*, **2013**, 6, 3380–3386
5. **H$_2$ and O$_2$ Evolution from Water Half-Splitting Reactions by Graphitic Carbon Nitride Materials.** A. Belen Jorge, David James Martin, Mandeep T. S. Dhanoa, Aisha S. Rahman, Neel Makwana, Junwang Tang, Andrea Sella, Furio Corà, Steven Firth, Jawwad A. Darr, and Paul F. McMillan. *The Journal of Physical Chemistry C*, **2013**, 117 (14), 7178–7185
6. **Conversion of solar energy to fuels by inorganic heterogeneous systems.** Kimfung Li, David James Martin and Junwang Tang. *Chinese Journal of Catalysis*, **2011**, 32 (6), 879–890
7. **CuO$_x$-TiO$_2$ junction: what is the active component for photocatalytic H$_2$ production?** Zhonlei Wang, Yuanxu Liu, David James Martin, Wendong Wang, Junwang Tang and Weixin Huang. *PCCP*, **2013**, 15, 14956–14960

Supervisor's Foreword

Solar water splitting using an inorganic semiconductor photocatalyst is viewed as one of the most exciting and environmentally friendly ways of producing clean renewable fuels such as hydrogen from abundant resources. Currently, there are many diverse semiconductors that have been developed, the majority for half reactions in the presence of sacrificial reagents. However, for industrial facilitation, there exists an essential, non-debatable trifecta of being robust, cheap and efficient for overall water splitting. To date, no system has combined all three, with most examples missing at least one of the necessary trio. Therefore one of the current challenges in the field is to develop low cost, highly efficient and stable photocatalysts for industrial scale-up use. In order to achieve that aim, researchers must focus on novel semiconductors to improve efficiencies and also understand the fundamental mechanisms.

In this thesis, Dr. David James Martin focuses on developing new photocatalysts for water photooxidation, reduction and overall water splitting. In doing so, the thesis aids to shed light on the mechanisms behind what makes certain photocatalysts either efficient or inefficient. Initially, the photooxidation of water using a novel faceted form of Ag_3PO_4 was investigated. A facile synthetic method was created that made it possible to control the exposing facets of silver phosphate in the absence of surfactants to yield tetrahedral crystals composed entirely of {111} facets. It was found that due to high surface energy of {111}, and low hole (h^+) mass in the <111> direction, Ag_3PO_4 tetrahedral crystals could outperform all other low index facets for the oxidation of water under visible light. The quantum yield was found to be nearly unity at 400 nm, and over 80 % at 500 nm. With the exception of Ag_3PO_4 tetrahedral crystals, no photocatalyst has exhibited quantum efficiencies reaching 100 % under visible irradiation. Therefore, the strategy of morphology control of a photocatalyst, led by DFT calculations of surface energy and charge carrier mobility, in order to boost photooxidation yield has been demonstrated to be very successful, and could be applied to improve other semiconductors in future research.

In parallel, hydrogen production from water was studied using the only known robust organic photocatalyst, graphitic carbon nitride (g-C$_3$N$_4$). It was discovered that using a reproducible preparation method, urea derived g-C$_3$N$_4$ can achieve a quantum yield of 26 % at 400 nm for hydrogen production from water; an order of magnitude greater than previously reported in the literature (3.75 %). The stark difference in activity is due to the polymerisation status, and consequently the surface protonation status as evidenced by XPS. As the surface protonation decreases, and polymerisation increases, leading to fast charge mobility and stronger reduction potential, thus an extremely high hydrogen production rate. The rate of hydrogen production with respect to BET-specific surface area was also found to be non-correlating; a juxtaposition of conventional photocatalysts whose activity is enhanced with larger surface areas—believed to be because of an increase in surface active sites.

Finally, overall water splitting has been demonstrated using Z-scheme systems comprising of a redox mediator, hydrogen evolution photocatalyst and oxygen evolution photocatalyst. Ag$_3$PO$_4$ was found to be not suitable for current Z-scheme systems, as it is unstable in the pH ranges required, and also reacts with both of the best known electron mediators used in Z-schemes, as evidenced by XRD, TEM and EDX studies. However, it has been demonstrated that urea derived g-C$_3$N$_4$ can participate in a Z-scheme system, when combined with either WO$_3$ or BiVO$_4$—the first example of its kind, resulting in a stable system for overall water splitting operated under both visible light irradiation and full arc irradiation. Further studies show water splitting rates are influenced by a combination of pH, concentration of redox mediator and mass ratio between photocatalysts. The solar-to-hydrogen conversion of the most efficient system was experimentally verified to be ca. 0.1 %. It is postulated that the surface properties of urea-derived graphitic carbon nitride are related to the adsorption of redox ions, however, further work is required to confirm these assumptions.

London   Prof. Junwang Tang
April 2015

Acknowledgments

First, I would like to thank my supervisor, Dr. Junwang Tang for his unparalleled guidance, expertise and insight throughout the entire project. Throughout the ups and downs, he remained focused and determined, two characteristics which have definitely rubbed off on me. He will remain a lifelong mentor and friend. I would also like to thank my second supervisor, Prof. Jawwad Darr, for many beneficial discussions, and for being very supportive in difficult times.

I am eternally grateful to Mark Turmaine and Jim Davy for helping me with SEM, Steve Firth for help with TEM/FTIR/Raman spectroscopy, Martin Vickers with XRD, and Rob Gruar for helping me with the most annoying instrument ever (ZetaSizer Nano). A special mention goes to Xiaowei Chen and the humble Juan Jose Delgado for consistently collaborating via their excellent TEM expertise.

A huge mention must go to Dr. Naoto Umezawa and Prof. Jinhua Ye (National Institute of Materials Science, 'NIMS', Japan). Professor Ye, who very kindly let me undertake a short research internship in her group, taught me in a very short space of time, to not only think outside the box, but also to make sure the box has the correct space group reflection conditions—so you know where the edges are! In my first international collaboration, Naoto was not only a good friend, but a patient and thoughtful man who had real faith that our work would be complementary and hence publishable. We had many discussions both personal and professional, and he will remain another lifelong friend. I thank them both for making my study in Japan thoroughly enjoyable.

I would also like to thank the forever enthusiastic and forward-thinking Dr. Stephen Shevlin, who was heavily involved in the second collaboration. His insight into DFT and TDDFT studies complimented and explained some of the experimental work I completed on carbon nitride. On a personal level Stephen also taught me about the ups and downs of reviewers, which I will never forget. Kaipei Qiu was also extremely helpful in contributing to a publication. Professor Z. Xiao Guo, who co-supervised the collaboration with Dr. Junwang Tang, was also precise, thoughtful and insightful throughout.

Dr. Albertus Handoko and Dr. Savio Moniz were two of the best postdocs a student could have. Before Albertus arrived, working without a PDRA was frankly a little difficult. Sometimes it is really beneficial to have somebody who can give you a quick answer, rather than search for it for hours. Definitely a friend for life, and despite him moving away from the group, I am sure our paths will cross again soon.

The people who made my Ph.D. actually fun in moments, kept me sane and helped with work during the later hours; Ben, Phil, Rhod, Seamus, Lawerence, Mayo, Amal, Chara, Noor, Mithila, Miggy, Eria, Moz, Toby, Jay, Vidal, Erik and all the rest of you guys.

To my fiancé Catariya, who was there throughout the good times and bad—I couldn't have done it without you. Finally, to all my family; Mum, Dad, brothers, Nan and Granddad—you literally kept me going and I will never forget the sacrifices made.

Contents

1 **Introduction: Fundamentals of Water Splitting
   and Literature Survey**... 1
   1.1 Fundamentals of Semiconductor Photoelectrochemistry.......... 3
       1.1.1 Semiconductor-Electrolyte Interface 3
       1.1.2 Charge Carrier Generation 5
       1.1.3 Photoelectrochemistry 7
       1.1.4 Photocatalytic Water Splitting 9
       1.1.5 Efficiency Calculations............................... 11
       1.1.6 Thermodynamic Limits 13
   1.2 Characterisation Methods for Photocatalysts.................. 15
       1.2.1 UV-Visible Spectroscopy 15
       1.2.2 Gas Chromatography 16
       1.2.3 Powder X-ray Diffraction (PXRD) 17
       1.2.4 Scanning and Transmission Electron Microscopy 18
       1.2.5 TGA-DSC-MS 19
       1.2.6 BET Method for Specific Surface Area
             Measurements....................................... 20
       1.2.7 Zeta Potential (ZP) Using Electrophoretic Light
             Scattering (ELS)................................... 21
       1.2.8 Attenuated Total Reflectance—Fourier Transform
             InfraRed (ATR-FTIR) Spectroscopy 22
       1.2.9 Raman Spectroscopy 23
       1.2.10 X-ray Photoelectron Spectroscopy (XPS) 25
       1.2.11 Elemental Analysis (EA) 26
   1.3 Literature Survey: Overview of Current Photocatalysts........ 27
       1.3.1 UV-Active Semiconductors 27
       1.3.2 Semiconductors Activated by Visible Light........... 29
       1.3.3 Semiconductors Activated by Visible Light........... 31
       1.3.4 Z-Scheme Systems................................... 37

       1.3.5   Effect of Morphology, Crystallinity and Size
               on the Activity of Photocatalysts   41
       1.3.6   Conclusions   43
   References   44

2   Experimental Development   55
   2.1   Reaction System   55
       2.1.1   Reactor   56
       2.1.2   Light Source   57
   2.2   Gas Chromatography: Selection and Calibration   58
       2.2.1   Gas Chromatography Setup   58
       2.2.2   Standard Gas and Calibration   60
   2.3   General Characterisation   62
       2.3.1   UV-Vis Spectrophotometry   62
       2.3.2   PXRD   62
       2.3.3   FE-SEM   63
       2.3.4   TEM   64
       2.3.5   BET Specific Surface Area   64
       2.3.6   ATR-FTIR Spectroscopy   64
       2.3.7   Raman Spectroscopy   64
       2.3.8   TGA-DSC-MS   65
       2.3.9   Zeta Potential (ZP) Measurements   65
       2.3.10 XPS   65
       2.3.11 Elemental Analysis   65
   References   66

3   Oxygen Evolving Photocatalyst Development   67
   3.1   Introduction   67
   3.2   Methodology   69
       3.2.1   Photocatalytic Analysis   69
       3.2.2   Synthesis Techniques   69
   3.3   Results and Discussion   72
       3.3.1   Initial Ag_3PO_4 Studies   72
       3.3.2   Facet Control of Ag_3PO_4 (Method 'D')   79
   3.4   Conclusions   91
   References   92

4   Hydrogen Evolving Photocatalyst Development   95
   4.1   Introduction   96
   4.2   Methodology   96
       4.2.1   Photocatalytic Analysis   96
       4.2.2   Synthesis Techniques   98
   4.3   Results and Discussion   98
   4.4   Conclusions 118
   References 120

5   Novel Z-Scheme Overall Water Splitting Systems 123
    5.1   Introduction 124
    5.2   Methodology 124
          5.2.1   Photocatalytic Analysis 124
          5.2.2   Synthesis Techniques 125
    5.3   Results and Discussion................................. 125
          5.3.1 Ag_3PO_4 Based Z-Scheme Water Splitting Systems 125
          5.3.2   Graphitic Carbon Nitride Based Z-Scheme Water
                  Splitting Systems 129
    5.4   Conclusions 141
    References... 142

6   Overall Conclusions and Future Work 145
    6.1   Overall Conclusions.................................... 145
    6.2   Future Work 147
    References... 149

List of Figures

Fig. 1.1   An example of an n-type semiconductor in contact with an electrolyte—**a** before equilibrium, **b** after equilibrium   4

Fig. 1.2   Idealised diagram of charge separation in a semiconductor photocatalyst.   5

Fig. 1.3   Band positions and magnitudes of the most well researched semiconductors. The *blue* region represents the redox minimum between 0 and 1.23 V versus NHE   6

Fig. 1.4   A typical PEC cell for photoelectrochemistry, with applied bias to drive proton reduction.   8

Fig. 1.5   An illustration of the three main processes in photocatalytic water splitting   9

Fig. 1.6   The maximum photoconversion efficiency possible as a function of the band gap wavelength (150 W Xe). Indicated with a *dashed line* is the maximum band gap wavelength at 610 nm. This corresponds with a minimum band gap of 2.03 eV for water splitting 14

Fig. 1.7   Schematic demonstration of a working ATR crystal in an ATR-FTIR spectrometer 23

Fig. 1.8   Raleigh and Stokes/Anti Stokes scattering processes in diagram form. Lowest energy vibrational state is n_0, and highest virtual states is v_2. *Arrows* represent direction of photon excitation (upwards) and relaxation (downwards) 24

Fig. 1.9   Formation of new valence and conduction bands by electron donor and acceptor atoms, enabling a UV-active material to respond to visible light 31

Fig. 1.10   Illustration of GaN:ZnO solid solution. Conduction band orbitals remain as Ga 4 s 4p, whilst valence band is hybridised to N2p, Zn3d and O2p 33

Fig. 1.11   Ball and stick (*left*), and molecular schematic drawing
            of g-C_3N_4 (*right*), using ChemSketch©. A single heptazine
            unit is highlighted in *red*. Graphitic carbon nitride
            is composed of; carbon (*grey*), nitrogen (*blue*), and hydrogen
            atoms (*white*)   36
Fig. 1.12   **a** Natural photosynthesis. **b** An artificial analogy involving
            two different photocatalysts possessing two different
            band positions (A is electron acceptor, D is electron
            donor)   38
Fig. 2.1   Schematic of proposed reactor **a** without magnetic stirring,
            b with stirring   56
Fig. 2.2   Borosilicate reactor for water splitting batch reactions   57
Fig. 2.3   Photograph of the gas chromatograph unit used during
            photocatalysis experiments.   58
Fig. 2.4   Idealised example chromatogram. A baseline is indicated
            by a horizontal *red* line, and integrated area under the curve
            (in μV min) is indicated by a shaded *blue* area   59
Fig. 2.5   GC area versus molar amount calibration curves
            for hydrogen (**a**), and oxygen (**b**). Equations of linear fits
            and R^2 values are noted in the *upper left* hand
            corner of the figure   61
Fig. 3.1   SEM micrographs of samples A1 (**a**, **b**), A2 (**c**, **d**),
            and A3 (**e**, **f**)   73
Fig. 3.2   **a** XRD patterns of samples *A1, A2* and *A3*. **a** UV-Vis
            absorbance spectra of samples *A1, A2* and *A3*   74
Fig. 3.3   Oxygen evolution under visible light ($l > 420$ nm)
            from water, in the presence of $AgNO_3$ (0.85 g),
            at neutral pH, using 0.2 g photocatalyst (*A1, A2, A3*)   74
Fig. 3.4   SEM micrographs of samples B1 (**a**, **b**), B2 (**c**, **d**),
            and B3 (**e**)   75
Fig. 3.5   XRD patterns for samples *B1, B2* and *B3*. The standard
            ISCD/JCPDS pattern for Ag_3PO_4 is shown as *red bars*.
            The Bragg peaks of the unknown phase are indicated
            by the symbol '⊗'.   76
Fig. 3.6   UV-Vis absorbance spectra for samples B1, B2 and B3   76
Fig. 3.7   Oxygen evolution under visible light ($l > 420$ nm)
            from water, in the presence of $AgNO_3$ (0.85 g),
            at neutral pH, using 0.2 g photocatalyst (*B1, B2, B3*)   76
Fig. 3.8   SEM micrographs of sample C1 in low (**a**)
            and high (**b**) resolution   77
Fig. 3.9   UV-Vis absorbance spectra of sample C1. The grey area
            visually demonstrates the range of the 420 nm long pass filter.
            A *black line* indicates the band edge and approximate
            absorption limit.   77

Fig. 3.10   XRD pattern for sample C1. The standard ISCD/JCPDS
            pattern for Ag_3PO_4 is shown as *red bars*, and the corresponding
            miller indices of each plane are indicated above each peak   78

Fig. 3.11   Oxygen evolution under visible light ($l > 420$ nm) from water,
            in the presence of $AgNO_3$ (0.85 g), at neutral pH, using 0.2 g
            photocatalyst (*C1*). Indicated are the regions of most
            importance; the initial period where O_2 evolution rate
            is high (rates are calculated from this point), and where
            the rate deviates from linearity in the later stages
            of the experiment   78

Fig. 3.12   Relaxed geometries for **a** {100}, **b** {110}, and **c** {111}
            surface of Ag_3PO_4 (calculations performed by Naoto Umezawa,
            NIMS).   80

Fig. 3.13   **a** Ag_3PO_4 $2 \times 2 \times 2$ supercell and **b** Ag_3PO_4 tetrahedron cell   81

Fig. 3.14   SEM micrographs of silver phosphate cubic crystals (**a**, **b**),
            and rhombic dodecahedral crystals (**c**, **d**).   81

Fig. 3.15   Low (**a**) and high (**b**, **c**) magnification SEM micrographs
            of Ag_3PO_4 tetrahedrons (20 cm^3 H_3PO_4). Low (**d**) and high
            (**e**) magnification SEM micrographs of Ag_3PO_4 tetrapods
            synthesised using 2 cm^3 H_3PO_4. **f** SEM micrographs
            illustrating the formation of Ag_3PO_4 tetrahedrons
            from tetrapods by varying the concentration of H_3PO_4
            (*left* to *right*; 2, 5, 10, 15, 20 cm^3 H_3PO_4).   82

Fig. 3.16   TEM micrograph tilt study, performed by collaborator,
            Xiaowei Chen. A tetrahedron indicated by *dotted lines*
            is rotated on axis from -66 to $+50°$ 83

Fig. 3.17   XRD pattern of faceted Ag_3PO_4 crystals   83

Fig. 3.18   UV-Vis absorbance spectra of faceted Ag_3PO_4 crystals   84

Fig. 3.19   Particle diameter histograms for Ag_3PO_4 **a** tetrahedrons,
            b cubes, **c** rhombic dodecahedrons.   86

Fig. 3.20   **a** Oxygen yield comparison of Ag_3PO_4 facets using a 300 W
            Xe light source under full arc irradiation, with $AgNO_3$ acting
            as an electron scavenger. **b** Oxygen yield comparison between
            tetrahedral Ag_3PO_4, previously reported random mixed
            faceted Ag_3PO_4 and $BiVO_4$ under 300 W Xe lamp full arc
            irradiation, using $AgNO_3$ as an electron scavenger. **c** Oxygen
            evolution of tetrahedral crystals using a 300 W Xe lamp fitted
            with 420 nm long pass filter, $AgNO_3$ was used as an electron
            scavenger (0.85 g). The mixed facet sample was synthesised
            using method C and Na_2HPO_4. $BiVO_4$ was synthesised
            according to a previously reported recipe.   87

Fig. 3.21   BET adsorption isotherms for Ag_3PO_4 tetrahedrons, cubes,
            and rhombic dodecahedrons   88

Fig. 3.22   XRD pattern before and after a full arc photocatalytic test
            for the photooxidation of water by tetrahedral Ag_3PO_4. $AgNO_3$
            was used as a scavenger, and thus Ag metal deposited
            on the surface (JCPDS 00-04-0783).   88

Fig. 3.23   **a** Oxygen evolution using a 400 nm band pass filter,
            and **b** Internal quantum yield variation using different band
            pass filters, using $AgNO_3$ as a scavenger, and a 300 W
            Xe light source   89

Fig. 4.1   XRD patterns of g-C_3N_4 synthesised using different precursors
            using 5 °C ramp rate, 600 °C, 4 h hold time   99

Fig. 4.2   ATR-FTIR spectra of powdered graphitic carbon nitride;
            annotated are characteristic bands for g-C_3N_4   99

Fig. 4.3   UV-Vis absorbance spectra for graphitic carbon nitride
            powders synthesized using different precursors at 600 °C. 100

Fig. 4.4   Raman spectra of different carbon nitride samples 101

Fig. 4.5   TEM micrographs of g-C_3N_4 synthesised from: **a** Urea,
            b thiourea, and **c** DCDA. The micrograph scheme at **d** shows
            a continual magnification of the g-C_3N_4 (urea) structure, and
            the resulting diffraction rings from microcrystalline regions. 101

Fig. 4.6   BET nitrogen adsorption/desorption isotherms. All samples
            were pre-prepared by calcination at 600 °C, for 4 h, with
            a ramp rate of 5 °C/min 102

Fig. 4.7   TGA-DSC-MS data from g-C_3N_4 synthesised from urea
            and thiourea 103

Fig. 4.8   TGA **a** and DSC **b** analysis of urea and thiourea. 104

Fig. 4.9   Hydrogen evolution using a 300 W Xe lamp, 3 wt% Pt,
            TEOA as a hole scavenger; **a** Full arc, **b** $\lambda \geq 395$ nm 106

Fig. 4.10   Hydrogen production with various synthesis parameters
            of urea derived g-C_3N_4 ($\lambda \geq 395$ nm, TEOA hole scavenger,
            300 W Xe lamp). **a** Urea based g-C_3N_4 synthesized at different
            temperatures, 4 h hold time (°C). **b** Urea synthesized
            at 600 °C using different ramping temperatures (°C/min).
            c g-C_3N_4 (urea) with different cocatalysts, 3 wt%. **d** Weight
            percent of Pt in comparison to photocatalyst (urea, 600 °C) 107

Fig. 4.11   XPS spectra from g-C_3N_4, for urea, DCDA and thiourea
            derived samples 109

Fig. 4.12   C1s and O1s XPS spectra of g-C_3N_4 synthesised from
            different precursors. 110

Fig. 4.13   Ratios of bonds within the N1s core level peak in different
            samples and their comparisons to the HER under visible light
            ($\lambda \geq 395$ nm), indicating decreasing proton concentration
            dramatically increases photocatalytic activity 110

Fig. 4.14   Zeta potential of g-C_3N_4 synthesised from different
            precursors (600 °C). 111

Fig. 4.15   Correlation between specific surface area normalised hydrogen
            evolution rate and bulk hydrogen content measured
            by elemental analysis   112
Fig. 4.16   Geometric and electronic structure of single sheet g-C$_3$N$_4$.
            a Supercell model of sheet carbon nitride, **b** supercell model
            of protonated carbon nitride. Nitrogen is denoted by *blue*
            spheres, carbon *brown* spheres, and hydrogen *light pink*
            spheres. **c** Total density of states for sheet carbon nitride
            (*black line*) and protonated carbon nitride (*red line*). Energy
            is with respect to the zero of the sheet carbon nitride
            simulation. The DOS of the protonated carbon nitride
            has been shifted so that the corresponding zero points align   113
Fig. 4.17   Total and site-decomposed DOS for protonated carbon
            nitride model. The strong contribution of the carbon atoms
            adjacent to the N3c site at the CBE is apparent, as the
            six atoms contribute almost as much
            to the DOS as the other twelve carbon atoms
            in the supercell   113
Fig. 4.18   Excited state properties of bare (sheet) and protonated
            carbon nitride; **a** optical properties of C$_{18}$N$_{28}$H$_{12}$ (*black line*)
            and C$_{18}$N$_{28}$H$_{13}$ (*red line*) clusters, the x-axis is labeled
            with respect to the vacuum level (E = 0) while dashed
            lines indicate the positions of the HOMO states of the two
            clusters, **b** lowest energy exciton of C$_{18}$N$_{28}$H$_{12}$ cluster,
            c lowest energy exciton of C$_{18}$N$_{28}$H$_{13}$ cluster. Orange isosurface
            indicates distribution of photohole upon photoexcitation, green
            isosurface indicates distribution of photoelectron upon
            photoexcitation. Isosurfaces were plotted at 0.005 |e|/Å^3.
            Blue spheres denote hydrogen atoms, grey spheres denote
            carbon atoms, white spheres denote hydrogen atoms.   114
Fig. 4.19   **a** Quantum yield of urea based g-C$_3$N$_4$; using band pass
            filters at specific wavelengths (*black line* represents
            absorbance, internal quantum yield is shown by *blue points*).
            b Stability test of the urea derived g-C$_3$N$_4$ under visible
            light irradiation ($\lambda \geq 395$ nm).   116
Fig. 4.20   **a** SEM micrograph of g-C$_3$N$_4$-MCA-DMSO-30-600. **b** UV-Vis
            absorbance spectra of g-C$_3$N$_4$-MCA-DMSO-30-600   117
Fig. 5.1   Powder X-ray diffraction data post water splitting
            in g-C$_3$N$_4$–Fe$^{2+/3+}$–Ag$_3$PO$_4$ system. Both main phase silver
            phosphate and sub phase silver sulphate are indicated
            by *red* and *blue* bars respectively   126
Fig. 5.2   **a** TEM micrographs of Ag$_3$PO$_4$ with AgI layer post water
            splitting, in g-C$_3$N$_4$–I$^-$/IO$_3^-$–Ag$_3$PO$_4$ system. Atomic spaces
            indicated by parallel lines. **b** Corresponding FFT of micrograph,
            diameter of diffraction spots are indicated   127

Fig. 5.3   **a** TEM micrograph of an Ag_3PO_4 crystal post water splitting,
             in $g\text{-}C_3N_4\text{–}I^-/IO_3^-\text{–}Ag_3PO_4$ system. Atomic plane distance
             indicated by parallel lines. **b** Corresponding FFT of micrograph,
             diameter of diffraction spot is indicated   127

Fig. 5.4   TEM micrographs (low res mode) combined with EDX
             of $Ag_3PO_4\text{-}AgI$ post water splitting in $g\text{-}C_3N_4\text{–}I^-/IO_3^-\text{–}Ag_3PO_4$
             system. **a** Full spectrum mapping of TEM-EDX micrograph
             ($Ag = green$, $O = pink$, $P = yellow$, $I = blue$). **b** Iodine
             mapping of TEM-EDX micrograph. **c** EDX spectrum
             of corresponding TEM micrograph area ($5 \times 5\ \mu m$), inset
             denotes weight % of map sum spectrum. *Note* Cu signal
             is from the copper/holey carbon grid used as a support,
             and other heavy elements such as Ca, Si, Co, Os (all 0 wt%)
             are artefacts.   128

Fig. 5.5   **a** Powder XRD pattern and **b** UV-Vis absorbance spectra
             of $g\text{-}C_3N_4$, WO_3 and $BiVO_4$ compounds prior to water
             splitting reactions   130

Fig. 5.6   SEM and TEM micrographs of different photocatalyst.
             a & **b** Commercial WO_3 crystals. **c** Synthesized $BiVO_4$
             crystals. **d** & **e** Synthesized $g\text{-}C_3N_4$ sheets   131

Fig. 5.7   Optimisation of water splitting rates using a $g\text{-}C_3N_4\text{–}BiVO_4$
             Z-scheme system with $FeCl_2$ as a redox mediator. **a** pH
             modification, **b** differing concentration of $FeCl_2$, **c** alternating
             the ratio of $g\text{-}C_3N_4$ to $BiVO_4$.   133

Fig. 5.8   Post water splitting XRD data of $g\text{-}C_3N_4$ //$BiVO_4$ system
             at different pH levels. Tetragonal phase BiOCl,
             JCPDS 00-006-0249, monoclinic phase $BiVO_4$,
             JCPDS 00-044-0081   134

Fig. 5.9   Stoichiometric water splitting (300 W Xe lamp source):
             a $g\text{-}C_3N_4$ (3 wt% Pt)$\text{–}FeCl_2\text{–}BiVO_4$ under full arc irradiation,
             pH 3, 1:1 photocatalyst weight ratio, 2 mM $FeCl_2$. **b** $g\text{-}C_3N_4$
             (3 wt% Pt)$\text{–}FeCl_2\text{–}BiVO_4$ under visible light irradiation
             ($\lambda > 395$ nm), pH 3, 1:1 photocatalyst weight ratio,
             2 mM $FeCl_2$.   134

Fig. 5.10   Water splitting using **a** $g\text{-}C_3N_4$ (3 wt% Pt)$\text{–}NaI\text{–}WO_3$
             (0.5 wt% Pt) under full arc irradiation, pH 8.3, 1:1
             photocatalyst weight ratio, 5 mM NaI and, **b** $g\text{-}C_3N_4$
             (3 wt% Pt)$\text{–}NaI\text{–}WO_3$ (0.5 wt% Pt) under visible light
             irradiation ($\lambda > 395$ nm), pH 8.3, 1:1 photocatalyst weight
             ratio, 5 mM NaI   135

Fig. 5.11   Optimisation of $g\text{-}C_3N_4\text{–}WO_3$ Z-scheme system with NaI as a
             redox mediator. **a** pH modification, **b** differing concentration
             of NaI, **c** alternating the ratio of $g\text{-}C_3N_4$ to WO_3   137

Fig. 5.12   Post XRD diffraction patterns from two different $g\text{-}C_3N_4$
             based water splitting systems   138

Fig. 5.13   BET isotherms of commercial WO_3 (SSA $= 1.15$ m^2g^{-1}),
            $BiVO_4$ (SSA $= 3.91$ m^2g^{-1}), and WO_3 platelets (2.45 m^2g^{-1}) 138
Fig. 5.14   Comparison of XRD diffraction patterns of WO_3 commercial
            crystals and as synthesised WO_3 platelets 139
Fig. 5.15   Low (**a**) and high (**b**) SEM micrographs of platelet WO_3 140
Fig. 5.16   **a** UV-Vis absorption spectra of commercial WO_3,
            and platelet WO_3 particles, Tauc plots of; WO_3 platelets
            (**b** & **c**), and WO_3 commercial particles (**c** & **d**)............... 141

List of Tables

Table 1.1 Summary of water splitting systems using a Z-scheme setup, currently reported in the literature........................ 40

Table 2.1 Gases and corresponding thermal conductivity at STP......... 59

Table 2.2 Area sampling data for H_2 and O_2 (0.5 cm^3) 62

Table 2.3 Data table for Fig. 2.5 63

Table 3.1 Summary of results on initial Ag_3PO_4 studies using various phosphate precursors.................................. 70

Table 3.2 Properties of different Ag_3PO_4 crystals 78

Table 3.3 Size, range, and standard deviation (S.D.) of various Ag_3PO_4 crystals.......................... 85

Table 4.1 Table showing the actual wt% of some metal precursors 97

Table 4.2 Summary of g-C_3N_4 properties synthesised from different precursors 106

Table 4.3 Ratios of bonds within the N1s core level peak in different samples and their comparisons to the hydrogen evolution activity 111

Table 4.4 Percentage breakdown of different bonds within the N1s spectrum..................................... 116

Table 4.5 Comparison of typical g-C_3N_4 photocatalysts reported for hydrogen production and the corresponding quantum yields 118

Table 5.1 AgI FFT diameter relation to a real distance................. 127

Table 5.2 Ag_3PO_4 FFT diameter relation to a real distance.............. 128

Table 5.3 Overall water splitting under full arc irradiation (300 W Xe lamp) using different redox mediated systems 132

Table 5.4 Intensity ratios of WO_3 PXRD Bragg peaks 139

Nomenclature

Acronyms and Initialisms

AM1.5	Air Mass 1.5
AQY	Apparent Quantum Yield
ATR-FTIR	Attenuated Total Reflectance Fourier Transform Infrared Spectroscopy
BCC	Body Centred Cubic
BET	Brenauer–Emmett–Teller
CB	Conduction Band
CBE	Conduction Band Edge
CCD	Charge-Coupled Device camera sensor
CHNX	Carbon Hydrogen Nitrogen Unknown (analysis)
DCDA	Dicyandiamide
DFT(+U)	Density Functional Theory (+ Hubbard term 'U')
DI	Deionised (water)
DMSO	Dimethyl Sulfoxide
DOS	Density of States
DSC	Differential Scanning Calorimetry
EA	Elemental Analysis
EDX	Energy Dispersive X-ray Spectroscopy
ELS	Electrophoretic Light Scattering
EM	Electromagnetic
EtOH	Ethanol
FFT	Fast Fourier Transform
FID	Flame Ionisation Detector
GC	Gas Chromatography
HER	Hydrogen Evolution Reaction
HOMO	Highest Occupied Molecular Orbital
ICSD	Inorganic Crystal Structure Database
IEP	Isoelectric Point

IQY	Internal Quantum Yield
IR	Infrared
JCPDS	Joint Committee on Powder Diffraction Standards
LED	Light Emitting Diode
LUMO	Lowest Unoccupied Molecular Orbital
MB	Methylene Blue
MCA	Melamine and Cyanuric Acid
MCA	Morphology Controlling Agent
MeOH	Methanol
MO	Methyl Orange
MS	Mass Spectroscopy
NADPH	Nicotinamide Adenine Dinucleotide Phosphate
NHE	Normal Hydrogen Electrode
OER	Oxygen Evolution Rate
PEC	Photoelectrochemical
PSI	Photosystem 1
PSII	Photosystem 2
PTFE	Polytetrafluoroethylene
PVP	Polyvinylpyrrolidone
PXRD	Powder X-ray Diffraction
QY	Quantum Yield
RhB	Rhodamine B (organic dye)
RHE	Reversible Hydrogen Electrode
RT	Retention Time
SCE	Saturated Calomel Electrode
SD	Standard Deviation
SEM	Scanning Electron Microscopy
SSA	Specific Surface Area
STH%	Solar-to-Hydrogen conversion efficiency
TAS	Transient Absorbance Spectroscopy
TCD	Thermal Conductivity Detector
TDDFT	Time Dependent Density Functional Theory
TEM	Transmission Electron Microscopy
TEOA	Triethanolamine
TGA	Thermo-Gravimetric Analysis
TOF	Turnover Frequency
TON	Turnover Number
UHV	Ultra-High Vacuum
UPS	Ultraviolet Photoelectron Spectroscopy
UV-Vis	Ultraviolet-Visible Spectroscopy
VB	Valence Band
VBE	Valence Band Edge
XPS	X-ray Photoelectron Spectroscopy
ZP	Zeta Potential

Roman Symbols

A	Absorbance (%)
A	Surface area
a	Mass of the powdered adsorbant
c	Speed of light (ms^{-1})
c	BET constant
d_{sc}	Space charge layer depth
E	Electric field
E_1	Heat of adsorption of first layer
E_{BG}	Band gap (eV)
E_{bulk}	Total energy of atomic bulk per unit
E_{CB}	Bottom of conduction band
E_F	Fermi level
E_γ	Energy of the incident x-ray
E_{KE}	Electron kinetic energy
E_L	Heat of adsorption for n layers > 1
E_s	Incident irradiance
E_{slab}	Total energy of atomic slab
E_{VB}	Top of valence band
e^-	Electron
F	Faraday constant
ΔG^0	Gibbs free energy change (kJ mol^{-1})
h	Planck's constant (J·s)
h^+	Hole
I_λ	Incident spectral photon flux
J_{BB}	Blackbody photon flux at wavelengths below that of the band gap wavelength
J_s	Absorbed photon flux
K	Scherrer shape factor
m_e^*	Electron mass
m_h^*	Hole mass
N	Number of atomic units in slab
N_A	Avogadro's number
n	Direct/Indirect band gap transition variable
n	Refractive index
n_y	Real vibrational mode 'y' (y $\geq$ 1)
P	Pressure (Pa)
p	Partial vapour pressure
p_0	Saturation pressure of the physisorbed gas
R	Reflectance (%)
ΔS_{mix}	Entropy of mixing
s	Adsorption cross section of the powder
T	Temperature (K)

T	Transmission (%)
U_{loss}	Energy lost per photon
V	Molar volume of gas
V_{FB}	Flat band potential
V_{ON}	Onset potential
V_{SCE}	Voltage of Saturated Calomel Electrode
V_{SSC}	Voltage of Silver-Silver Chloride Electrode
v	Particle velocity
v_a	Volume of gas adsorbed at standard temperature and pressure
v_m	Adsorbed gas quantity per volume to produce a monolayer
v_x	Virtual state 'x' (x $\geq$ 1)

Greek Symbols

α	Absorption coefficient
β	Line broadening at half the maximum intensity
$\Delta \varphi_{sc}$	Space charge layer potential
δ_p	Penetration depth
ε_0	Permittivity of free space
ε_ρ	Permittivity of the suspension medium
η	Viscosity of the suspension medium
η_c	Photoconversion efficiency
θ	Angle
λ_0	Wavelength of incident light in vacuum
λ_{BG}	Band gap wavelength
μ	Electrophoretic mobility
ν	Frequency (s^{-1})
ν_δ	Doppler shift frequency
σ	Standard Deviation
τ	Mean size of the ordered crystalline domains
Φ	Diameter (mm)
Φ_{loss}	Radiative quantum yield
$\Phi_{loss\,P}$	Proportion of absorbed photons converted into conduction band electrons
ϕ	Instrument work function

Chapter 1
Introduction: Fundamentals of Water Splitting and Literature Survey

In response to noticeable climate change and growing energy demands, governments across the globe have pledged to drastically change the way they produce energy [1]. One of the most promising solutions to these two concerns is solar energy conversion. The 3,850,000 EJ of solar radiation which the Earth absorbs in 1 year, dwarfs our current energy consumption (around 474 EJ per year). Harnessing this power and using it efficiently would provide us with enough clean, renewable energy, to last us indefinitely.

Electrolysis of water is a conventionally employed and well-researched process. Hydrogen can be evolved from this reaction; however, it requires an enormous amount of energy. In order to bypass this energy input, but still maintain the production of hydrogen, the use of solar radiation is employed. Solar energy can be used to split water in three ways. Firstly, solar concentrators and catalysts are able to thermally decompose water, but this takes place at high temperatures, requiring materials to be stable at high temperature. More widely used is the second method, using photovoltaic devices to photoelectrolyse water. However currently, electrolysers and photovoltaic systems are reasonably costly [2].

The third method is water photolysis; using specific wavelengths of radiation to promote charge carrier separation in a semiconductor; enabling water splitting through a photogenerated voltage. This option provides many advantages which the previous two do not. Many inorganic semiconductors are simply metal oxides/nitrides/sulphides; which are often non-toxic, cheap, and abundant. Furthermore, since the semiconductor photocatalyst houses both charge separation and dissociation reactions, losses are expected to be considerably low [3]. The fuel produced from the photochemical water splitting is termed "solar hydrogen" and can be used directly in, for example, fuel cells. Although a common view is that hydrogen is dangerous to transport due to its high flammability, it has been reported that moving hydrogen through pipelines proves more efficient than electricity [4], which could prove to be of critical importance in the future. Another advantage of hydrogen fuel is that when used, in either direct combustion or a fuel cell stack, the resulting 'waste' is merely water—making the whole cycle essentially carbon free.

© Springer International Publishing Switzerland 2015   1
D.J. Martin, *Investigation into High Efficiency Visible Light Photocatalysts for Water Reduction and Oxidation*, Springer Theses,
DOI 10.1007/978-3-319-18488-3_1

The semiconductor used in water photolysis will serve the same function as the light reaction centre in the natural photosynthetic reaction. However, the system's energy vector will not be ADP or NADPH, but H_2 (solar hydrogen). As the water is oxidised, and oxygen is released, protons are reduced to hydrogen:

General reaction for water splitting

$$2H_2O + 4h^+ \rightarrow O_2 + 4H^+$$
$$4H^+ + 4e^- \rightarrow 2H_2$$

(1.1)

"Solar hydrogen" can also be utilised as a chemical feedstock, to produce ammonia, in upgrading (hydrocracking) fossil fuels, producing methanol, and high purity acids amongst other chemicals. Solar hydrogen therefore could directly replace current feedstocks which are obtained via fossil fuels, reducing CO_2 emissions [5].

Historically, photoelectrochemical water splitting was first reported in 1972, on a TiO_2 electrode under ultra-violet (UV) radiation. However, UV light represents merely 4 % of solar radiation that hits the Earth's surface [6]. Considering this, and the fact that visible light constitutes over 44 % of radiation incident on the Earth's surface, the development of visible light-driven photocatalysts has become the major focus of researchers in the field [7–9]. Regarding the performance, a generally accepted target of 10 % "Solar-to-Hydrogen conversion efficiency" (STH%) is required for commercial viability [10]. Yet this isn't the only requirement; a semiconductor photocatalyst must also be robust and stable enough to last for an extended period of time, and furthermore be cheap enough to be cost effective.

Since there are over 200 tested materials to date, and many different well-documented approaches for fabricating each one, it is crucial to separate the types of compounds that are promising for further research, from the ones that are not. Pure water splitting on one photocatalyst is extremely difficult, both in terms of thermodynamics and charge carrier kinetics, therefore a route often pursued by academics is the investigation of half reactions. Half reactions can selectively produce usable fuels such as molecular H_2 or O_2, at the expense of an electron donor or acceptor, which itself is spent. Certain donors/acceptors can also be successively reduced and oxidised ('redox mediators'), and can be used in conjunction with two photocatalysts forming a Z-Scheme for overall water splitting, utilising two simultaneous half reactions that are spatially and temporally separated. Consequently extensive studies been carried out on compounds that can produce either hydrogen or oxygen separately in scavenger systems, such as TiO_2 and $SrTiO_3$ [11], which have relatively wide band gaps, and also into materials with narrow band gaps, such as WO_3 [12], GaAs [13] and CdS [14]. The typical trend in these reports is that wide band gap materials have low maximum photoconversion efficiencies, but have a high durability throughout a range of conditions. Narrow band gap semiconductors have the ability to absorb a wider portion of the electromagnetic (EM) spectrum, giving them higher theoretical maximum photoconversion efficiencies [15]. However, narrow band gap semiconductors often have band positions which inhibit the mutual evolution of oxygen and hydrogen, and can result in a photocorrosion effect, where redox reactions occur on the semiconductor itself instead of reducing/oxidising water [16].

Numerous compounds have been found to be active in the visible part of the spectrum; however, they have not provided a high enough efficiency at a low enough cost and the stability for commercialisation [7, 17, 18]. The search for newer visible light active photocatalysts, or further advancing the newer generation of photocatalysts, is now the current focus of the community.

In the research field, there are two main ways to study the activity of a photocatalyst. The first is to construct a photoelectrochemical (PEC) cell, which uses electrical bias to drive photoelectrons around a circuit, and thus generate a photocurrent. This is analogous to hydrogen production in the second method (colloidal reaction), which is to directly place the semiconductor photocatalyst in water, and monitor hydrogen and oxygen evolution using gas chromatography. A PEC cell relies on the efficiency of a photocatalyst, it also requires the refining of an extra component—the boundary between semiconductor and electrode/circuit (discussed later in Sect. 1.1.1). Logically then, the cheapest and most facile way of screening different photocatalysts is in a colloidal setup, whereby the photocatalyst is immersed in water in a sealed reactor.

Therefore considering this strategy, the overall project goal was to utilise the simplicity of a colloidal set up in order to screen for highly efficient and novel visible light active photocatalysts, creating a materials platform for future water splitting systems to be built on. Furthermore, there were five main objectives that would enable the overall project aim to be achieved. The first was a thorough exploration and analysis into the design of semiconductor photocatalysts in the form of a comprehensive literature survey. It considered the robustness, cost, relative abundance of constituent elements, and finally the cleanliness of synthesis methods used in manufacturing photocatalysts.

The second was development of an efficient semiconductor photocatalyst for photocatalytic water oxidation using visible light. The third goal was to manufacture a highly active material for the production of molecular hydrogen from water under visible light irradiation. The fourth major aim in terms of experimental validation was to construct an overall water splitting system based on success and findings of the previous objectives. Ultimately, the synthesised compounds were studied using a variety of characterisation and analytical techniques in order to divulge a mechanistic insight into rate determining steps and factors which influence activity would be discovered.

1.1 Fundamentals of Semiconductor Photoelectrochemistry

1.1.1 Semiconductor-Electrolyte Interface

An equilibrium process takes place when a semiconductor is immersed in a redox electrolyte due to the disparate nature of the Fermi level; a flow of charge between the two phases ensues, forming a Helmholtz double layer [19]. The junction between the phases is akin to that of a Schottky barrier in a semiconductor-metal interface [16]. The Helmholtz double layer potential drop is accompanied by a further potential drop within the electrolyte at low ion concentrations in the Gouy layer. As the charge carrier density in a semiconductor is often rather small, and the Debye length is very large, a space charge region can be formed. This region dictates the electrode potential: the difference between semiconductor and electrolyte. The potential of the space charge layer ($\Delta\varphi_{sc}$) is shifted downwards for a p-type semiconductor, and upwards for an n-type semiconductor (Fig. 1.1), this is "band bending".

For an n-type semiconductor, there are excess electrons in the bulk (more negative Fermi level), therefore in order to equilibrate, electrons flow from the surface of the semiconductor into the electrolyte, leaving a positive charge behind, the electrode potential.

The Fermi level in an intrinsic semiconductor should lie in the centre of the band gap (equal number of electrons and holes); doping with cations or anions shifts the Fermi level respectively to more positive potentials, and more negative potentials. As shown in Fig. 1.1, the Fermi level in the semiconductor experiences a much larger shift than that of the one in the electrolyte solution. This is due to the lack of energy states in the semiconductor, excluding defect sites in the band gap region (in the forms of donor and acceptor states), in comparison to that of the electrolyte. A curious phenomenon which arises specifically in nanostructures is the diminished or complete absence of band bending. This occurs when then particle size is less than or equal to the depth of the space charge layer. This means the nanostructures can be unpredictable, demonstrating different behaviour depending

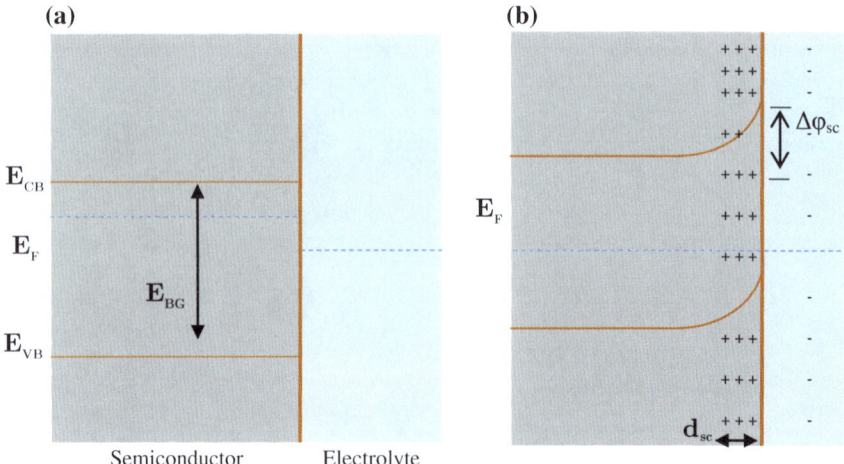

Fig. 1.1 An example of an n-type semiconductor in contact with an electrolyte—**a** before equilibrium, **b** after equilibrium. Adapted from Refs. [15, 20]. Symbols are as follows: E_{CB} bottom of a conduction band, E_{VB} top of a valence band, E_{BG} band gap, E_F Fermi level, d_{sc} depth of space charge region

on particle size and shape [21]. Electron-hole pairs induce electric fields which counteract band bending, and in turn raise the Fermi level towards the conduction band [22].

1.1.2 Charge Carrier Generation

Electrons located in the valence band (VB) of a semiconductor can be excited by photons which have energy greater than the band gap of semiconductor—the energy difference between the top of the filled valence band and the bottom of the empty conduction band (CB). The photoexcited electrons are promoted from the top of the VB to the bottom of the CB, simultaneously leaving positively charged holes in the VB (Fig. 1.2). If the electron-hole pairs (exciton) escape recombination they can travel to the surface of semiconductor and act as redox species for the splitting of water, organic decomposition, or the reduction of CO_2 [23].

In order to reduce protons to hydrogen fuel, electrons in the CB are required to have a reduction potential that is more negative than the redox potential of H^+/H_2, 0 eV versus Normal Hydrogen Electrode (NHE). Water oxidation occurs when the hole potential is more positive than the redox potential of O_2/H_2O, $+1.23$ eV versus NHE. Therefore, a minimum theoretical band gap of 1.23 eV is required.

This corresponds with the Gibbs free energy change (ΔG^0) for overall water dissociation:

$$H_2O + 2e^- \rightarrow H_2 + 0.5O_2 = 237\,kJ\,mol^{-1} \tag{1.2}$$

$$\Delta G^0 = -nFE^0 \tag{1.3}$$

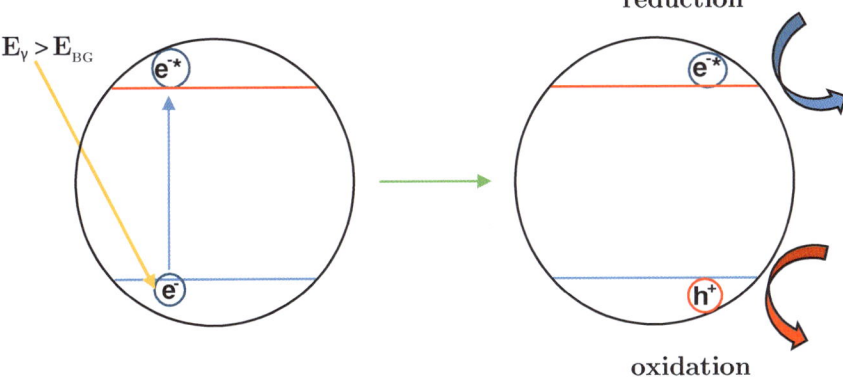

Fig. 1.2 Idealised diagram of charge separation in a semiconductor photocatalyst

where n is the number of electrons per mole product (2 electrons in this instance), F is the Faraday constant (units: C mol^{-1}) and E^0 is the electrode potential;

$$\Delta G^0 = -2 \cdot \left(9.64 \times 10^4\right) \cdot 1.23$$

$$\Delta G^0 = 2.46\,eV \times \left(9.64 \times 10^4\right)$$

$$\Delta G^0 = (-)237\,kJ\,mol^{-1}$$

In essence the minimum practical energy required to drive a photocatalytic reaction is much higher due to energy losses associated with the over-potentials required for the two chemical reactions and the driving force for charge carrier transportation. Studies disagree on this minimum limit, but according to reports it can be as low as 1.6 eV [24] and as high as 2.5 eV [25]. Therefore the reaction in itself is thermodynamically difficult. Since ultraviolet (UV) light ($\leq$400 nm) constitutes less than 4 % of the solar spectrum, and 43 % of light is in the visible region (400–750 nm), development of a photocatalyst with activity from the UV through to visible wavelengths (400–750 nm) and with a VB and CB that straddle the reaction potentials is one of key goals in obtaining optimum solar energy to fuel efficiency [26]. The band gaps of some simple semiconductors have been measured by various authors and are shown below (Fig. 1.3). As can be seen from Fig. 1.3, the general trend is that oxide semiconductors have very deep (positive vs. NHE) valence bands, facilitating stable oxidation reactions, but limiting light absorption. Conversely non-oxide semiconductors having smaller band gaps and more negative band positions, yet can suffer from photocorrosion.

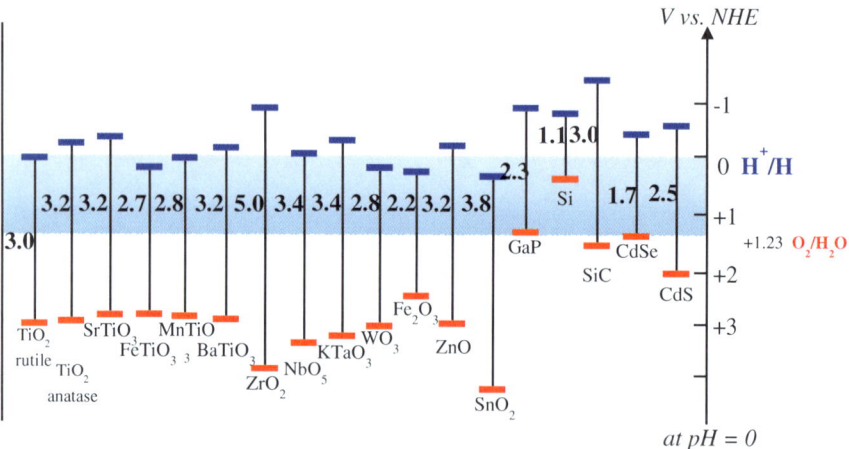

Fig. 1.3 Band positions and magnitudes of the most well researched semiconductors. The *blue* region represents the redox minimum between 0 and 1.23 V versus NHE. Inspired by Jing et al. [26]

1.1.3 Photoelectrochemistry

The first demonstration of conventional water splitting as we know it used a titania anode and a platinum cathode, and the reaction was driven by an electrical bias [6], therefore the system was in fact a photoelectrochemical cell. In modern photoelectrochemical cells (PEC cells), there are three electrodes, immersed in a redox electrolyte. The semiconductor photocatalyst acts as the working electrode, and the counter electrode is often platinum gauze, which due to the tiny overpotential is perfect for reducing protons to hydrogen fuel. The modern reference electrode is either Ag/AgCl (Silver-Silver Chloride—denoted SSC) or saturated calomel (Saturated Calomel Electrode—SCE), but historically, a normal or reversible hydrogen electrode (NHE/RHE) was used. They have been replaced due to their lack of practicality. However, they are often referenced and converted accordingly: $V_{SSC} = +0.197$ V versus NHE, $V_{SCE} = +0.242$ V versus NHE [27].

When an external bias is applied to the system, it is necessary to measure the current produced from the counter to the working electrode. The reference electrode should draw no current at all. This applied bias has the role of changing the Fermi levels of both the semiconductor and solution, often resulting in two different Fermi levels [15, 27, 28]. When there is no external bias applied to the system, and no band bending, the potential of the Fermi level is termed the flatband potential (V_{FB}). Here there is no photocurrent since all charge carriers recombine. If the flatband potential is high enough, it can be approximated to be the onset potential (V_{ON}) for the photocurrent [22]. Both flatband potentials of the CB and VB are determined intrinsically by the material, but are affected slightly by the pH of the electrolyte [28].

One of the main reasons behind performing photoelectrochemistry is not only to study and monitor the photocurrent, but to also drive the reduction of protons and produce hydrogen fuel. If the semiconductor's conduction band edge is fractionally anodic in comparison with the H_2 redox potential, then a voltage must be applied externally to the system in order to reduce the photogenerated electrons and turn H^+ into H_2 (Fig. 1.4).

Using photoelectrochemistry, it is possible to split up the overall reaction for water splitting into different parts, which are shown below [29].

Equation 1.1. General reaction for water splitting

$$H_2O \rightarrow H_2 + 0.5O_2$$

In basic solution:

Basic anodic reaction

$$2h^+ + 2OH^- \rightleftarrows 0.5O_2 + H_2O \qquad (1.4)$$

Basic cathodic reaction

$$2e^- + 2H_2O \rightleftharpoons H_2 + 2OH^- \qquad (1.5)$$

Fig. 1.4 A typical PEC cell for photoelectrochemistry, with applied bias to drive proton reduction. Inspired by Kudo and Miseki [23]

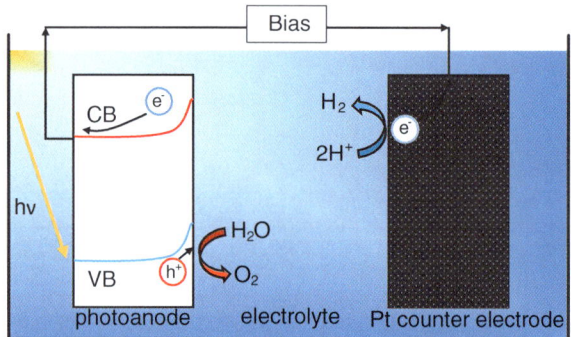

In acidic solution:

    Acidic anodic reaction

$$2h^+ + H_2O \rightleftharpoons 0.5O_2 + 2H^+ \tag{1.6}$$

Acidic cathodic reaction

$$2e^- + 2H^+ \rightleftharpoons H_2 \tag{1.7}$$

Note: these equations are for an n-type semiconductor, where the semiconductor is the anode. If the semiconductor is p-type, then it acts as the cathode.

Another side note on photoelectrochemical reactions has been noted by Rajeshwar regarding specific terminology [25]. The author mentions that the term "photoelectrochemical" should describe any reaction where light is used to enhance the electrochemical process. This thermodynamic process could be either "uphill" (positive ΔG^0—photosynthetic) or downhill (negative ΔG^0—photocatalytic). The term photoelectrolysis can be used for a photoelectrode in an electrochemical cell. Photocatalysis should be applied for those systems which are suspensions or colloidal, and require no external bias (but still with a positive ΔG). The phrase photoassisted water splitting should be used to describe experimental setups where the reaction process is a product of both photon energy and that supplied by bias.

One of the main problems with a photoelectrochemical setup is that the connection between the semiconductor photoanode and the driving bias must be almost perfect in order to gain a true insight into the activity of the photocatalyst. For example, the semiconductor could be theoretically efficient, but if the connection is poor, electrons cannot be shuttled to the counter electrode, and recombine with holes. Therefore the complexity of the device adds to the difficulty and hence cost of making a solar-to-fuel system, which is a major problem in the field. Often, a thin semiconductor film is grown onto a conducting substrate, with the aim that the penetration depth of light does not exceed the thickness of the film. For wavelengths 200–1000 nm, the penetration depth in most semiconductors (δ_p, the distance at which the intensity of incident electromagnetic radiation falls to 1/e or about ~37 % of its original value in a given material) increases with larger wavelengths, a direct consequence of band-gap absorption. That is to say most blue light is absorbed at the surface, whilst red light penetrates deep into the bulk. This becomes crucial in semiconductors which are used as photocatalysts because photogenerated charge carriers must migrate to the surface in order to participate in redox reactions. Therefore in thin films, excitons created deep in the bulk (in photocatalysts responding to visible light) are less likely to generate a photocurrent and split water. Along with poor film adhesion to the conductive substrate, it often leads to poor photocurrent in devices [30].

1.1.4 Photocatalytic Water Splitting

The previous section focused on using a semiconductor simply as a working electrode, and having the proton-electron recombination centre—the counter electrode, as a separate entity. There are many studies and reviews however which focus on colloidal systems; suspensions of photocatalyst particles within a glass batch reactor immersed in an electrolyte [7, 31, 32]. No external bias is applied, so the particles act naturally as a functioning anode and cathode (Fig. 1.5). This is

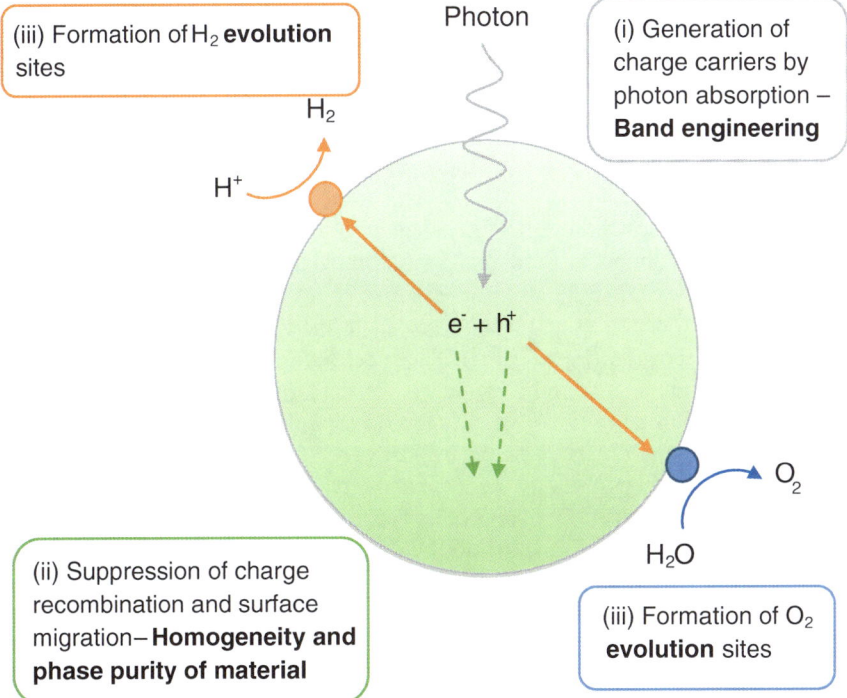

Fig. 1.5 An illustration of the three main processes in photocatalytic water splitting. Adapted from Ref. [23]

often a good test of a semiconductors' photocatalytic ability; one can see if hydrogen and oxygen can be evolved without bias, which is the eventual goal of idealised water splitting. It is arguably thought that powdered/colloidal catalyst systems would be ideal for large scale industrial hydrogen generation purely because of its straightforward nature [23]. There are no electrical connections needed, and if a good scavenger is used, an accurate efficiency can be obtained. However, some studies argue that colloidal photocatalyst systems have more use as a simple and effective method for testing the efficiency of different compounds.

Having hydrogen and oxygen together in a contained atmosphere is potentially explosive. Due to this fact, safety protocols should be followed when using a suspension system and an efficient photocatalyst for pure water splitting (e.g. using inert atmospheres and small volume reactors). Water splitting using colloidal systems on a larger industrial scale however might be impractical, as flammability limits might be reached unless the system is made more complex by using compressors and separating units to effectively shuttle hydrogen and oxygen mixtures away from reactors. However for lab-scale materials testing, suspension systems are an effective tool in particular for a half reactions. For instance, if a compound is found to be highly efficient, it could then be integrated into a PEC cell system, and hydrogen and oxygen would be produced in separate compartments of one cell.

Electron acceptors and donors (vernacular; electron and hole 'scavengers', respectively) are compounds which have a very high relative electronegativity (for electron acceptors/scavengers) or electropositivity (for electron donors/hole scavengers), and can therefore stop the production of either hydrogen (electron donor) or oxygen (electron acceptor) by combining with the respective charge carrier [33]. Scavenger compounds must also have a high solubility in water in order to donate or accept charge readily. Electron acceptors include $S_2O_8{}^{2-}$, $Ce(SO_4)_2$, $FeCl_3$ but the most frequently used is Ag^+ from $AgNO_3$, as it has been shown to scavenge electrons more efficiently than any other electron acceptor [34, 35]. Electron donors commonly used are: ethanol, methanol, triethanolamine (TEOA), Na_2S, $Na_2S_2O_3$ and Na_2SO_3.

The overall water splitting reaction process can be considered to contain three main parts (as shown in Fig. 1.5): (i) charge carrier (electron/hole) promotion following absorption of a photon, (ii) charge carrier separation and migration, (iii) redox reaction between surface species and the charge carriers. The first two steps are interrelated, and are both photophysical processes, whilst the final step is a purely chemical process. A total photocatalytic reaction is hence a complex combination of photophysical and photochemical processes. Besides research being carried out regarding enhancing the ability of visible light harvesting in a semiconductor (part (i)), there is also considerable development in the water splitting field with view to develop a rapid charge separation system (part (ii)). It has also standard practice to experiment with the material's particle morphology and alter the surface structure modification to increase the reaction rate of redox couples (part (iii)).

There have been many examples within semiconductor literature (which will be mentioned later in this chapter) that are able to selectively produce hydrogen or oxygen from water using charge carrier scavengers. Unfortunately, most are unable to evolve both gases in their correct stoichiometric ratio due to electron/hole recombination and lack of active sites with low overpotential on the surface. This leads to the obvious conclusion that although the correct positioning and energy of the band gap are important criteria, there are other factors which govern photocatalytic activity. Even if the hole and electron potential are sufficient to thermodynamically split water, they will recombine if active sites at the surface are not present. 'Cocatalysts', as they are termed, are usually loaded on the surface of the semiconductor to act as active sites for pure water splitting (NiO), H_2 evolution (Pt, Au, Ru) or oxygen evolution (RuO_2). This is because many oxide based semiconductors do not possess a conduction band which is high enough to reduce protons to H_2 without catalytic help. Most crucially, cocatalysts also act as electron or hole sinks, and prevent the recombination of charge. By localising electron and holes at specific catalytic sites for a much longer time period, the probability of a charge carrier participating in a redox reaction greatly increases. One report even suggests the double loading of cocatalysts; which improves the performance even further [9]. The in situ photodeposition of cocatalysts was demonstrated by Bamwenda, and shown to be the most effective method of loading a cocatalyst— in terms of activity [36]. This is due to the catalyst being deposited directly and selectively on an active site, where it is most effective. Without cocatalyst loading, factors such as charge carrier recombination—fuelled by kinetic competition, may occur at a superior rate than the necessary surface redox reactions, which can be detrimental to the reactivity in an overall water splitting reaction. However, there are exceptions; metal oxides for the most part do not need an oxidation catalyst due to the correct positioning and composition of oxygen 2p orbitals. With the exception of a few photocatalysts (ZnO for example), oxides are resistant to photocorrosion.

1.1.5 Efficiency Calculations

Many methods and equations can be used to calculate the efficiency of a semiconductors ability to decompose water. However, several excellent review papers have been published, [15, 23, 27] with proposed guidelines to conform to when performing efficiency analysis. In this section, the most important equations from these papers will be discussed, and it will be divulged which of these are suitable for analysis.

Commonly, the production of hydrogen or oxygen is given with the unit micromole of product per hour, per gram of photocatalyst; 'μmol h^{-1} g^{-1}'. Despite being acceptable unit for presenting rates, it is unsuitable for determining efficiency, since most gas evolution rates depend on the experimental conditions: light source, reactor size, amount of catalyst, and type of Gas Chromatograph (GC) amongst others. This makes evolution rates incomparable to some extent. The number of incident photons can be calculated using a thermopile or silicon detector/photodiode. Unfortunately, the number of absorbed photons (by the photocatalyst) is often considerably less than that of incident light due to scattering effects and transmission. But, the amount of photocatalyst used in an experiment should still be thought about carefully; whereby the catalyst concentration is adjusted to account for variable dispersivity, and minimise reflective losses. Therefore, it is possible to use a trial and error process, whereby catalyst concentration is varied, and then compared to the product produced in 'μmol h^{-1} g^{-1}'. As demonstrated by Li Puma, it is also possible to model the ideal catalyst concentration in a reactor by optimising the optical thickness of a semiconductor [37]. However, to simplify matters in this study, all scattered (externally reflected) photons by semiconductor particles and reactor walls, are not taken into account herein. The quantum yield obtained should actually be termed 'Apparent Quantum Yield' (AQY) where incident photon flux is measured outside the reactor. A more accurate measurement of the number of photons incident on the photocatalyst surface can be performed by subtracting the flux from behind the reactor, from the flux measured in front of the reactor. The quantum yield in that case is termed 'internal quantum yield' (IQY). The number of reacted electrons can be calculated by using the amount of evolved hydrogen or oxygen [38]. This can be understood as the percentage of photons which cause charge carriers to become reactive.

Internal Quantum yield

$$IQY(\%) = \frac{Number\ of\ reacted\ electorns}{Number\ of\ incident\ photons} \times 100 \qquad (1.8)$$

The Turn-Over Number (TON, Eq. 1.9) is used to illustrate whether a process is truly catalytic or not. Successfully used in homogenous catalysis, the TON can also be applied to heterogeneous catalysis with some minor modifications. It is normally defined as the number of reacted molecules per active site, but the active site does depend on the type of reaction (reduction/oxidation) [38]. Sometimes the number of active sites can be very roughly approximated to the number of molecules in the photocatalyst, though this grossly underestimates the actual turnover number.

Turnover Number

$$TON = \frac{Number\ of\ reacted\ electrons}{Number\ of\ active\ sites\ on\ photocatalyst} \qquad (1.9)$$

QY and TON are different than Solar-To-Hydrogen conversion efficiency, STH%, which is defined as:

STH%

$$STH(\%) = \frac{Output\ energy\ as\ H_2}{Energy\ of\ incident\ solar\ light} \times 100 \qquad (1.10)$$

A worthy note is that although the STH% equation only takes into account 'output energy as H$_2$', it cannot be used for half-reactions (H$_2$ production with a hole scavenger), and must be only applied to overall water splitting reactions, such as a Z-scheme. This is because the scavenger is consumed, and the process does not solely rely on water. That is to say that in the reformation of methanol (as an example hole scavenger), hydrogen is produced not only from water but from methanol itself. Redox mediators in Z-scheme reactions (such as $Fe^{2+/3+}$ and I^-/IO_3^-) aren't exhausted/consumed, and therefore Z-scheme systems can be evaluated using STH%.

Besides efficiency calculations, it is often noteworthy to examine the data using basic stoichiometry. In an overall water splitting reaction, hydrogen and oxygen should evolve in an approximate 2:1 ratio in the absence of an electron donor/ acceptor. If the amounts are not in a stoichiometric ratio, then it is clear that the reaction isn't proceeding photocatalytically; other factors could be coming into play (leaking, degradation etc.). During oxygen evolution, Ag^+ is reduced to Ag^0 on the surface of some particles (from electron scavenger $AgNO_3$). This leads to the deactivation of the photocatalyst as Ag metallic species block light absorption.

1.1.6 Thermodynamic Limits

In Sect. 1.1.2, it was mentioned that due to energy losses associated with overpotentials, the theoretical minimum band gap of a semiconductor required to split water was not 1.23 eV, but much higher. Murphy et al. investigated the overpotentials which cause the band gap to be increased; their most important findings are presented in this section [15]. Murphy et al. also note that their findings are in agreement with other groups; Weber and Dignam and Bolton et al. [39–41].

In an ideal semiconductor, all photons with energy greater than E_{BG} would be absorbed. Valence band electrons can earn promotion to the conduction band, and then they are transferred to the electrolyte, where a pair of such electrons can lead to a hydrogen molecule being produced. The absorbed photon flux would be:

Absorbed photon flux (J_s)

$$J_s = \int_0^{\lambda_{BG}} I_\lambda(\lambda)d\lambda \tag{1.11}$$

where λ_{BG} is the band gap wavelength $(\lambda_{BG} = hc/E_{BG})$ and I_λ is the incident spectral photon flux $(m^{-2} \ nm^{-1} \ s^{-1})$. One can then form an upper limit to the photoconversion efficiency:

Limits of photoconversion efficiency (η_c)

$$\eta_C = \begin{cases} \frac{J_S \Delta G^0(1-\Phi_{loss})}{E_S} & if \ E_{BG} \geq \Delta G^0 + U_{loss} \\ 0 & if \ E_{BG} \leq \Delta G^0 + U_{loss} \end{cases} \tag{1.12}$$

The term 'Φ_{loss}' accounts for the radiative quantum yield, which is more specifically the ratio of re-radiated photons to absorbed photons. Therefore, the collective term '$1 - \Phi_{loss}$' is the amount of absorbed photons which have led to conduction band electrons being produced. The re-radiation of photons is due to blackbody radiation from the excited state. The value of $\Phi_{loss \ P}$ that relates a maximum value of the efficiency is given by:

Proportion of absorbed photons converted into conduction band electrons (Φ_{lossP}):

$$\Phi_{loss\,P} \approx 1/ln\left(J_s/J_{BB}\right) \tag{1.13}$$

where J_{BB} is the blackbody photon flux at wavelengths below that of the band gap wavelength, λ_{BG}. $\Phi_{loss\,P}$ has been calculated to be small, less than 0.02 eV for the range of band gap wavelengths that photocatalysts lie in, and for all intents and purposes, is independent of the spectrum of incident radiation.

In Eq. 1.12, the term U_{loss} is the energy lost per photon; which turns out to be a significant loss term, unlike $\lambda_{loss\,P}$. This enables a criterion to be set up for the lower limit of the band gap energy of a semiconductor that can be used for water splitting. The loss is composed of two main parts; a thermodynamic part and a kinetic part. The thermodynamic part is equal to $T\Delta S_{mix}$, where T is temperature and ΔS_{mix} is the entropy of mixing which governs the production of conduction band electrons by photon absorption ($\geq$0.4 eV). Combined with the second contribution, which arises from kinetic losses due to the overpotentials for oxygen and hydrogen ($\geq$0.4 eV), the estimation for U_{loss} is $\geq$0.8 eV. This therefore leaves the minimum band gap energy at 2.03 eV, which means that the limiting band gap wavelength is 610 nm (as shown in Fig. 1.6).

Murphy et al. reviewed the effect on photoconversion efficiencies caused by the different light sources [15]; as one can see from Fig. 1.6, using different light sources yields different results. The conclusion drawn is that the spectra of artificial light sources, such as Xenon, mercury, and xenon-mercury lamps, do not accurately represent the spectrum of solar irradiation. This leads to quoted efficiencies being higher than those thermodynamically possible for photocatalysts. Despite this, arc lamps are the most accurate representation of the solar spectrum available for most scientific research groups.

The air mass spectra (AM) are the most accurate way of replicating the spectrum of solar radiation found on the Earth's surface [15]. The AM spectra account for the amount of the Earth's atmosphere radiation has to go through, and also,

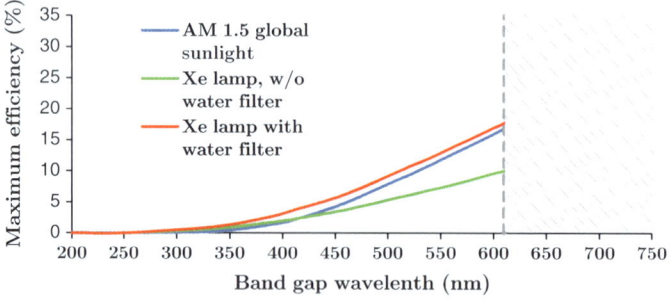

Fig. 1.6 The maximum photoconversion efficiency possible as a function of the band gap wavelength (150 W Xe). Indicated with a *dashed line* is the maximum band gap wavelength at 610 nm. This corresponds with a minimum band gap of 2.03 eV for water splitting. Reproduced from Ref. [15]

what radiation is filtered out in the process. An AM 1.5 filter removes some UV and IR radiation, like our own atmosphere, and also takes into account the angle at which the sun's radiation passes through. AM 1.5 corresponds to 48.2°, derived from AM = 1/cosα (α is the zenith angle of the sun), relative to α = 0, with the illuminated surface inclined at 37°, leaving an 11.2° incidence angle. The two types of AM spectra are direct and global. Direct is just from a single á angle, while global takes into account scattered light from all angles.

Murphy et al. concluded that at short wavelengths a xenon lamp with no water filter has a larger spectral flux in comparison to the AM 1.5 global spectrum (in the real world the atmosphere absorbs most harmful UV radiation at $\lambda \leq 300$ nm). Furthermore, older xenon lamps have decreased UV flux in comparison with newer ones. One last point, which can be seen from Fig. 1.6 is that the xenon lamp without a water filter can yield overestimated efficiencies at $\lambda \leq 410$ nm; this means that wide band gap materials are often overestimated, and narrow band gap materials are underestimated.

It is often worthwhile to also conduct experiments with, and without a long pass filter, especially for visible light activated photocatalysts. Whilst no filter illuminates the sample with both UV and visible light, long pass filters block out all UV light ($\lambda \geq 400$ or 420 nm). Therefore a comparison can be drawn between the contribution of UV or visible photons on the photocatalytic activity of a compound. This is especially helpful with new or doped visible light photocatalysts, where one might want to see if the main contribution comes from UV or visible photons. A classic example of this is nitrogen doped TiO_2; the absorption spectra shows an absorption tail into the visible range, however, the compound is unable to evolve molecular oxygen under visible light, even in the presence of scavengers [42]. Therefore the contribution of visible photons to water splitting activity is almost zero, although the compound can degrade some organic molecules.

1.2 Characterisation Methods for Photocatalysts

When engineering either new compounds, or attempting to modify existing photocatalysts, it is extremely important to classify them correctly. A wide variety of characterisation methods are available at present, and these not only help to confirm results, but also aid in understanding the inner workings of a photocatalyst.

1.2.1 UV-Visible Spectroscopy

Calculating the absorption spectra for a material is invaluable, it enables analysis of the wavelength range where a material absorbs, and also how strong. UV-Visible spectrophotometers ('UV-Vis') probe through the electromagnetic spectrum from ultraviolet through to the visible range. Molecules often undergo electron transitions at these frequencies, since semiconductors have discrete energy levels corresponding to allowed electron states. Therefore one can determine the size and type of band gap (direct or indirect). This is done using a Tauc plot, often used in thin films [43]:

Tauc relation

$$(\alpha h\nu) = A(h\nu - E_{BG})^n \tag{1.14}$$

where α is the absorption coefficient, $h\nu$ is the energy of the incident radiation, A is the constant which is based on effective masses of electrons and holes, and n can take the values of 0.5 and 2, for a direct or indirect band gap transition.

Often, UV-Vis spectrophotometers will give data where the wavelength is on the abscissa and either the absorbance (A) or reflectance (R) on the ordinate. The absorption coefficient can be calculated using:

Kubelka-Mulk relation

$$\alpha = \frac{(1-R)^2}{2R} \tag{1.15}$$

The reflectance is related to the absorption by the following equations:

Transmission-reflectance-absorption relation

$$1 = T + R + A \tag{1.16}$$

where T is the transmission in %. Often spectrophotometers will give the absorbance as a function of a logarithm. If this is to be converted to a percentage then the following relation can be used:

Absorbance conversion

$$A(\%) = 1 - 10^{-A\log 10} \tag{1.17}$$

The band gap of semiconductor is approximately equal to the absorption edge, and so one can also find this from extrapolating towards the x-axis on a plot of wavelength (x-axis) versus absorption ($\log_{10}$ y-axis), similar to a Tauc plot (Eq. 1.14). Then using the following relation, calculate the band gap:

Planck relation (nm)

$$E = \frac{hc}{\lambda} \rightarrow h\nu = \frac{1240}{\lambda} \tag{1.18}$$

The absorption edge can be blue shifted by decreasing the particle size, and is often attributed to either a quantum size effect or a direct charge-transfer transition. If the band gap region possess a long tail, and appears to be red shifted, then this would conclude that there are additional sub band gap states present [44].

1.2.2 Gas Chromatography

The process of water splitting often yields two gases—oxygen and hydrogen, and being able to calculate the amount of each produced is vital. Gas chromatography is used to analyse concentrations and amounts of gas in a given atmosphere, often to a great resolution. A sample of gas is injected; the elements of the gas (mobile phase) are separated by a stationary phase 'column' (often stainless steel tubing, densely packed with diatomaceous earth), compounds elute at different times and pass through the detector where a signal is generated and data can be produced.

Gas Chromatographs (GC's) are usually fitted with either a Thermal Conductivity Detector (TCD), or a Flame Ionisation Detector (FID). The most suitable detector for simultaneous hydrogen and oxygen detection would be a TCD. As the name suggests, the TCD detects changes in thermal conductivity produced by the sample from a standard gas flow (carrier gas), originally using a wheatstone bridge and resistors which present varying heat loss depending on analyte. Standard carrier gases are nitrogen, helium and argon. Argon is ideal for hydrogen and oxygen detection since it has a thermal conductivity much less than the respective gases [45]. Helium could be used, however due to its similar thermal conductivity to hydrogen, it would be unfavourable, and also yield negative peaks. Furthermore, in a photocatalysis setup, it would be useful to monitor nitrogen levels in the system. Primarily, nitrogen levels can be monitored to check for leaks in the system, but also to track whether a nitride/oxynitrides photocatalyst is degrading or not (nitrogen levels will increase in both cases).

There are two types of analysis associated with GC's; qualitative and quantitative. Data is represented as a graph of retention time (x-axis) and detector response (y-axis); the retention time is specific to each compound (qualitative), and the detector response is dependant (with a TCD) on the thermal conductivity and amount of each element (quantitative). Chromatographic data therefore yields a spectrum of 'peaks' separated only by time. The area under each peak represents the amount of substance present. By using a calibration curve (responses from a series of known concentrations/amounts), one can then determine the amount of substance present, and correctly identify the element/compound by its distinct retention time.

1.2.3 Powder X-ray Diffraction (PXRD)

PXRD is used to quickly classify the atomic and molecular structure of a crystalline material, including phase, purity and composition. X-rays are fired at an (unknown) compound, causing elastic (Raleigh) scattering. Given angles of coherent and incoherent scattering are related to the lattice parameters in reciprocal space by Bragg's law [46]:

$$n\lambda = 2d \sin \theta$$

where n is an integer, λ is the wavelength of electromagnetic radiation, d is the spacing between atomic planes (not atoms) and θ is the angle between atomic planes and incident X-ray angles.

By varying the angle of both the X-ray source and the detector (2θ), it is possible to attain a relationship between intensity and scattering angle, termed a diffraction pattern. This pattern is then run through a database to quickly identify the element or compound according intensities at set scattering angles. This is an especially good technique for gaining insight into photocatalytic mechanisms, since band gaps can change with phase and crystallinity, and it is also very easy to see impurities, both of which can affect photocatalytic activity.

Further analysis can also yield more specific features of compounds, such as crystallite size, and the average distance between molecules. If there is a weak peak response, in conjunction with a broad width, then this is indication of poor crystallinity. However, good peak responses with broad features can often mean the particle size is very varied. The average particle or crystallite size can be calculated using Scherrer's equation [47]:

Scherrer's equation

$$\tau = \frac{K\lambda}{\beta \cos\theta} \tag{1.19}$$

where K is the shape factor (near 1, but crystallite shape dependant), λ is the X-ray wavelength, β is the line broadening at half the maximum intensity, θ is the Bragg angle, and τ is the mean size of the ordered crystalline domains, which is approximately the grain size. As a side note, the Scherrer equation is only applicable to nano-sized particles, and is limited to a grain size of 0.1 μm. More importantly, the Scherrer formula actually calculates a lower bound to the particle size. This is because the width of the diffraction peak can be determined by many different factors, not just crystallite size, such as inhomogeneous strain and instrumental effects [48]. Scherrer's equation should not be used where access to more appropriate particle sizing techniques (such as SEM or TEM) are available.

1.2.4 Scanning and Transmission Electron Microscopy

Scanning Electron Microscopy (SEM) and Transmission Electron Microscopy (TEM) are two different techniques used for imaging the surface structure of semiconductors. SEM is often used to identify the size of particles and examine the morphology of the surface, especially with samples which are reasonably thick. The SEM produces a high energy electron beam that interacts with the surface of the sample, which in turn emits secondary electrons, backscattered and diffracted backscattered electrons, all of which are captured by detectors and are transformed into a high depth of field image. This gives the image a 3-dimensional appearance despite being a two dimensional image.

TEM involves the transmission of an electron beam through a thin sample (~100 nm); the beam interacts with the sample and refracts around atoms, generating an image on a photographic film or CCD. Used mainly in imaging thin films or nanoparticles which are less than 10 nm, it can also be used to detect nano-scale surface dopants such as metallic/crystalline elements, which often to not show on SEM image. Unlike SEM, TEM allows the examination of non-conductive sampling. If a non-conductive sample is studied using SEM, then an effect known as 'charging' occurs, often ruining the sample, and making imaging very difficult. If the CCD or photographic film is in the correct position, then it is also possible to see diffraction rings or spots from the sample, indicative of crystalline nature. If certain instrumental parameters are known, then it is possible to correlate the size of these rings with known d-spacing of the crystal structure (obtained from XRD). Then, using a diffraction pattern, it is possible to identify which crystal plane is being observed, which is especially useful when studying faceted crystals.

Scanning and Transmission Electron Microscopes often come with an additional feature: EDX (Energy Dispersive X-ray Spectroscopy, sometimes abbreviated to EDS). This analytical technique is often used for elemental analysis of samples, resulting in a spectrum of elements and their respective ratios. The electron beam from the SEM is used to stimulate ground state electrons within the sample, forcing them to be ejected. The empty ground state left behind is filled by an electron from a higher shell, and an X-ray is emitted. The emitted X-ray has energy which is equal to the difference between the two shells—which is characteristic of individual elements [49]. In photocatalysis studies, this can be used to identify the distribution of cocatalysts on the surface, or look at impurities.

1.2.5 TGA-DSC-MS

Thermo Gravimetric Analysis—Differential Scanning Calorimetry—Mass Spectroscopy (TGA-DSC-MS), is complementary set of techniques which couple thermal analysis and evolved gas analysis, essentially observing characteristic changes that a compound undergoes under varying temperature. It is particularly useful for investigating phase changes and condensation reactions of polymers. TGA is used to monitor mass of a solid or liquid with temperature, and with a high precision, measure mass difference due to evaporation or sublimation. This can then be plotted on a simple x-y curve; mass loss (as a percentage) versus temperature or time. DSC gives information on thermal transitions of compounds—whether endothermic or exothermic reactions occur at a specific temperature. Again, a simple x-y curve of sample heat change in comparison to a reference, versus temperature or time. MS ionises compounds from the TGA-DSC gas-solid-fragment stream and then measures their mass to charge ratio; culminating in spectra of atomic masses and their relative concentration at a given time. One can then bring the spectra together in time or temperature, and plot what masses (and thus elements or compounds) are present at that instant, and in what concentration.

1.2.6 BET Method for Specific Surface
Area Measurements [50]

The Brauner-Emmett-Teller (BET) theory describes the physisorption process of gas molecules on a solid surface, resulting from the relatively weak Van de Waals forces. The theory can be extended to a method for measuring the specific surface area of materials—especially relevant for photocatalysts, since a greater surface area can potentially lead to more active sites per unit area, leading to an increase in activity. Adapted from Langmuir theory, which describes the monolayer absorption of a gas on a solid, BET theory extends the concept by allowing for multilayer adsorption, and is calculated by measuring the following equation:
   The BET equation

$$\frac{1}{v_a\left[\left(\frac{p_0}{p}\right)-1\right]} = \frac{(c-1)}{v_m c}\left(\frac{p}{p_0}\right) + \frac{1}{v_m c} \tag{1.20}$$

where v_a is the volume of gas adsorbed at standard temperature and pressure (STP), p and p_0 are the partial vapour pressure and saturation pressure of the physisorbed gas (often nitrogen, N_2) at ~195.5 °C, c is the BET constant, v_m is the adsorbed gas quantity per volume to produce a monolayer.
   The BET constant (c).

$$c = e^{\frac{(E_1-E_L)}{RT}} \tag{1.21}$$

where E_1 is the first layer's heat of adsorption, and E_L is the corresponding heat of adsorption for n layers >1.
   Equation 1.21 can be plotted simply as a 'y = mx + c' graph, the result termed a 'BET plot'. The ordinate term is $\frac{1}{v_a\left[\left(\frac{p_0}{p}\right)-1\right]}$, whilst $\frac{p}{p_0}$ can be plotted on the abscissa. Linearity is found at approximately $0.01 \le \left(\frac{p}{p_0}\right) \le 0.3$ and thus extrapolating from this point using the gradient A to y-intercept I gives v_m and c according to the equations:

$$v_m = \frac{1}{A+I} \tag{1.22}$$

$$c = 1 + \frac{A}{I} \tag{1.23}$$

The BET specific surface area (SSA) is then given by:
   BET SSA equation

$$S_{BET} = a\left(\frac{v_m N_A s}{V}\right) \tag{1.24}$$

where a is the mass of the powdered adsorbant, N_A is Avogadro's number, s is the adsorption cross section of the powder, and V is the molar volume of gas.

From a practical point of view it is in fact much simpler to attain a value for specific surface area, since software accompanies the surface area analyser which can calculate all of the above automatically. All that is required is a precise measurement of the mass of the sample, and to correctly pick the point of linearity. Linearity (a line of best fit) can be improved by a constant trial and error process which minimises the correlation coefficient (R-factor), so that the value is as close to 1 as possible (acceptable values are normally ≥ 0.999).

1.2.7 Zeta Potential (ZP) Using Electrophoretic Light Scattering (ELS) [51]

In a colloid, microscopic particles are suspended in a solvent by nature of their ZP; the potential difference between the solvent and a fixed layer of solvent attached to the dispersed particle. The electrokinetic properties are determined by the electric charge distribution at the 'double layer' surrounding the particle. When immersed in an electrolyte such as water, an ionic particle becomes surrounded with counter ions, with a charge opposite to that at the surface. The first layer is formed from chemical interactions, whilst the second layer that comprises the double layer is attracted by nature of the Coulomb force, and thus electrically screens the first layer. The second layer (slipping plane) is only lightly bound to the particle and is made of free ions, and thus its depth can be influenced by the amount of positive or negative charge within the solution. The potential at the slipping plane is the zeta potential, as measured from the surface of the particle.

In short, zeta potential determines the stability of the colloid, as highly charged particles repel each other, and prevent flocculation. As a rule of thumb, a colloid with a ZP of 0 mV (Isoelectric Point, IEP) will undergo rapid coagulation or flocculation, ~40 mV will demonstrate good stability, and ≥ 60 mV exhibits excellent stability.

A zeta potential measurement not only provides information of the stability of a colloid, but upon further analysis, can give relative information about the surface charge—whether positive or negative, and relative magnitude. Its application to photocatalysis stems really from the latter point; the information gained regarding surface charge can help to explain phenomena experienced in synthesis methods, and more importantly, differences in activity due to surface species such as H^+ and OH^-.

One effective way of measuring zeta potential is to use micro-electrophoresis (rate of particle movement under an electric field) in conjunction with electrophoretic light scattering. The particle-electrolyte colloid is placed in a specialised cuvette-type cell, an oscillating electric field is then applied, which causes the particles to move with a velocity proportional to their zeta potential. In order to measure this velocity, a laser beam is passed through the cell, and the Doppler shift frequency observed is proportional to the dispersed particle's mobility according to:

$$v_D = \mu \frac{nE}{\lambda_0} \sin \theta \tag{1.25}$$

where v_d is the Doppler shift frequency, n is the refractive index of the medium, λ_0 is the wavelength of incident light in vacuum, and θ is the scattering angle.

The electrophoretic mobility μ is defined as:

$$\mu = \frac{v}{E} \tag{1.26}$$

where v is the velocity of the particles, and E is the electric field applied.

Electrophoretic mobility is directly proportional to zeta potential:

$$\mu = \frac{\varepsilon_0 \varepsilon_r \zeta}{\eta} \tag{1.27}$$

where ε_0 is the permittivity of free space, ε_r is the permittivity of the suspension medium, and η is the viscosity of the suspension medium.

Therefore by measuring velocity, under a known electric field, and having known constants (often pre-programmed into the apparatus), one can attain the zeta potential.

1.2.8 Attenuated Total Reflectance—Fourier Transform InfraRed (ATR-FTIR) Spectroscopy [52]

'ATR-FTIR' is a fast infra-red spectroscopic technique for probing the local surface structure of a material, with minimal preparation. From an analysis point of view, infrared spectroscopy enables the detection and fingerprinting of covalent bonds. This is useful for both characterisation and analysis of impurities/contaminants on the surface of semiconductor photocatalysts. Since most photocatalysts are crystals with simple molecular structures, they possess few IR-active bonds. However, organic contaminants or impurities are intrinsically more molecularly complex, and thus will easily be detectable using IR spectroscopy.

In IR spectroscopy, a sample is illuminated with a monochromatic light source. Absorption will occur if the frequency of light corresponds to the vibrational frequency of the bond; this is then apparent the in the spectrum of transmittance, and can be correlated to how much energy has been absorbed. FTIR is more counter-intuitive, yet has a much faster sampling time and higher signal to noise ratio [53]. In FTIR, a broadband/polychromatic light source (all frequencies to be measured) is used to illuminate the source. However, the beam first passes through a Michelson interferometer, a beam splitting device which causes a deliberate path length shift (termed 'retardation'), subsequently illuminating the detector after hitting the sample with a mixed interference beam of all wavelengths. Beam retardation is caused by a moving mirror within the interferometer, which has real units of length. Therefore the detector produces an interferogram; a fixed set of retardation values and a corresponding set of intensities. A Fast Fourier Transform (FFT) is automatically applied by the instrument, converting time/space to frequency. The resultant FTIR spectrum is a set of inverse lengths (cm^{-1} is often used) versus intensity. Bonds can thus be assigned to intensities at certain energies (wavenumbers).

ATR-FTIR is similar in respect to standard FTIR methods, however it has considerable advantages in terms of preparation method, and is now gaining a considerable hold on the academic market. A solid or liquid can simply be placed on the ATR crystal with a small amount of pressure to form an intimate contact, and then analysed. Compared to traditional FTIR methods, which rely on transmission and therefore a diluted sample/thin film, ATR-FTIR relies on Total Internal Reflectance (TIR) and thus does not have this issue. The sample is also then reusable, and uncontaminated. The beam of infrared light is incident on the ATR crystal; passing through and reflecting off the sample (see Fig. 1.7). As a result of TIR, an evanescent wave is set up, propagating through the crystal and reflecting of both crystal and sample consecutively [54]. As long as the ATR crystal has a higher refractive index than the sample, the technique is viable. The penetration depth into the sample is approximately 0.5–2 μm.

1.2.9 Raman Spectroscopy

Complementary to FTIR spectroscopy in the sense that it can provide a compounds fingerprint, Raman spectroscopy is less widely used method of detecting vibrational, rotational and other low frequency modes in a compound [55]. In FTIR spectroscopy, a polychromatic light source is used, and detection of activated IR bonds depends on the loss of a certain frequency due to absorption. In Raman spectroscopy, a monochromatic source illuminates the sample, which is in turn scattered by the sample; it is not an absorption technique. An incident photon polarises the electron cloud and creates an inherently unstable excited virtual state, which is then stabilised by reemitting another photon of similar, but not identical energy. Figure 1.8 illustrates this process. A degree of electron cloud deformation

Fig. 1.7 Schematic demonstration of a working ATR crystal in an ATR-FTIR spectrometer [52]

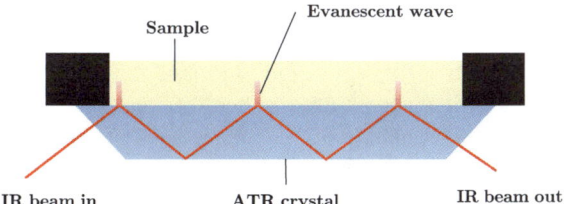

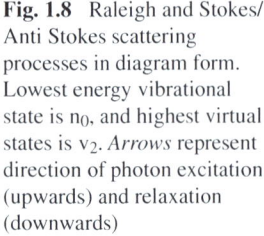
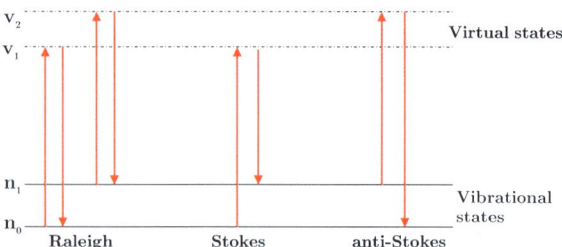

Fig. 1.8 Raleigh and Stokes/ Anti Stokes scattering processes in diagram form. Lowest energy vibrational state is n_0, and highest virtual states is v_2. *Arrows* represent direction of photon excitation (upwards) and relaxation (downwards)

(polarizability) caused by the photon with regards to the vibrational coordinate is necessary for a molecule to undergo a Raman effect. The magnitude of polarizability is proportional to the Raman scattering intensity. The dependence on the polarizability is thus the difference between Raman and IR spectroscopy, as IR spectroscopy only probes molecules whereby photons interact with the molecular dipole moment. This enables Raman spectroscopy to analyse vibrational/rotational transitions which are effectively not allowed as determined by the rule of mutual exclusion in centrosymmetric molecules [56].

Elastic/Raleigh scattering occurs when the photon excites a ground state vibrational mode to a virtual state and subsequently returns to the same energy ground state. Common Raleigh scattering is around 10^6–10^8 more prevalent than Raman scattering, which is a very weak process, and thus needs a monochromatic source.

Raman scattering can be divided into two different types of scattering; 'Stokes' and 'Anti Stokes'. 'Stokes' scattering arises when a photon excites a ground state vibrational mode (n_0) to an initial virtual state (v_1), and is then reemitted with a lower energy, the difference in photon energy is thus the difference between the vibrational energy states ($n_1 - n_0$). Some molecules however may be in an excited state (from thermal energy), thus 'Anti Stokes' scattering takes place when the incident photon excites a higher energy vibrational state (n_1) to a higher virtual state (v_2), which then returns to the ground state (n_0). The energy of the virtual states are determined by the power of the incident photon.

Practically, Raman spectroscopy is relatively quick, there is little sample preparation (powdered samples are simply put in the way of a laser beam), and is non selective. The monochromatic laser light is Raman scattered by the sample, collected by a lens and focused onto a detector. Photons which are Raleigh scattered (i.e. no energy change) are rejected by use of a notch or edge filter. The software within the instrument then compares the initial laser wavelength (λ_0) with the Raman scattered wavelength (λ) and calculates the Raman shift ($\lambda_0 - \lambda$). The generated plot is thus a function of Raman shift (cm^{-1}) versus intensity, and shows what vibrational/rotational bonds are present in the molecule/sample.

In terms of its applicability to analysing semiconductor photocatalysts, Raman spectroscopy can be used in two different ways. The first is simply to detect possible contaminants on the surface of a semiconductor photocatalyst, and correlate with photocatalytic activity. Often, contaminants act as either recombination traps or electron/hole scavengers, and having a detrimental effect on photocatalytic activity by preventing the reduction or oxidation of water. Secondly, since Raman spectroscopy probes local molecular structure through the Raman shift, a change in Raman shift at a particular wavelength would infer that the bond is either moving up or down in terms of energy. It could therefore be interpreted that an increase or decrease in bond energy results in a distortion in local structure (i.e. change in electron orbital overlap), which can result in either a decrease or increase in photocatalytic activity, as demonstrated with $BiVO_4$ by Yu and Kudo [31].

1.2.10 X-ray Photoelectron Spectroscopy (XPS)

XPS enables the characterisation of the electron energy distribution within a compound by irradiation with an X-ray source [57]. Primarily a qualitative surface sensitive elemental analysis technique, XPS is capable of measuring atomic composition (%), chemical and electronic state, and bonding configurations of elements in a sample, all whilst being completely non-destructive. Depth profiling deeper into the bulk is possible using ion etching, but native use is limited up to approximately 10 nm into the bulk from the surface [58]. Electrons in shells up to but not including the valence band can be excited by X-rays, whilst valence electrons are excited by UV light, and measured using a similar technique, Ultraviolet Photoelectron Spectroscopy (UPS). Both techniques must be under Ultra High Vacuum (UHV) in order for the electrons to reach the detector. Most elements can be quantitatively and qualitatively measured, with the exception of hydrogen and helium—the diameter of the orbitals are too small therefore the catch probability is close to zero [59].

When a sample is irradiated with a high energy X-ray source (ca. 1.4 keV for a common aluminium K-alpha source), core level electrons are excited to the vacuum level and are then detected by an electron energy analyser. The binding energy (E_b) of the core level is then calculated using the following formula:

Conservation of energy

$$E_b = E_\gamma - (E_{KE} + \phi) \tag{1.28}$$

where E_g is the energy of the incident X-ray, E_{KE} is the measured energy of the electron measured by the instrument, and ϕ is the work function of the instrument.

As incident electrons hit the detector, both electron intensity and binding energy are recorded as a peak-like spectrum, often with a sizeable background. Since the core level binding energies in an atom are unique to each element, it is initially possible to identify the element and electron orbital. Further peak analysis can then separate each individual contribution to that element, for example, a carbon 1 s peak might have convolutions of 2 or 3 different bonds in different binding energies; C-C, C-O or O-C = O [57].

XPS has considerable applicability in the characterisation of photocatalysts, evidenced by its importance in recent high profile articles [60–62]. Since numerous UV-active photocatalysts are cheap, robust, and relatively efficient (discussed in 1.3), many research groups have actively tried to extend their absorption into the visible region so that more light can be absorbed and increase the potential STH%. One of the ways is to use dopants—ions which alter the band structure and effectively insert a sub-band gap state to enable more light absorption. Monitoring the atomic percentage of these dopants after preparation is useful in order to ascertain the optimum dopant percent—this can effectively be done with XPS. In addition, using XPS depth profiling combined with traditional electron orbital labelling, it is possible to differentiate between surface species and bulk-bound ions through location mapping and oxidation state identification [63].

1.2.11 Elemental Analysis (EA)

EA, or in the most common and applicable form, CHNX analysis, is the quantitative and qualitative determination of the elemental composition and weight of a compound by combustion. Often used to determine carbon, hydrogen, and nitrogen, in organic molecules, other compounds can also be detected using additional gas separation methods. Put simply, a compound of unknown composition is combusted in an oxygen rich atmosphere, the constituents separated by gas chromatography, and in modern apparatus, analysed using a high temperature TCD [64, 65].

A small amount (ca. 10 mg) of sample is weighed in a small aluminium capsule. The Al capsule is then moved to a high temperature furnace (in some cases, over 1800 °C) and combusted under a pure static oxygen atmosphere. An extra purge of oxygen is then added at the end to facilitate complete combustion. The produced gases are then passed through reagents that produce more stable gases such as CO_2, H_2O and N_2. This step also removes phosphorus, sulphur and halogen based compounds. In order to 'scrub' oxygen and oxides, the gases then pass over a copper section, which reduces oxides of nitrogen to elemental nitrogen. The gas is then homogenized at a constant temperature and pressure when passed through a mixing volume chamber.

The gas is then passed through a series of TCDs, with H_2O traps in between. The difference between the two signals as a loss of water is thus proportional to the hydrogen concentration in the sample being analysed. Another set of TCDs with a CO_2 trap in between calculate the carbon concentration, and finally nitrogen is measured using a helium reference over a TCD. Other compounds, such as sulphur and oxygen can be analysed using specialised reaction tubes.

The atomic weight of C, H or N is useful for analysis of semiconductor elements containing the three elements. In particular, the composition of graphitic carbon nitride (one of the only organic semiconductor photocatalysts) can be analysed, and the carbon to nitrogen ratio calculated.

1.3 Literature Survey: Overview of Current Photocatalysts [66]

This subsection draws on relevant publications in the field, ranging from well-documented literature, to newer, novel findings and alternative approaches to achieving photocatalytic water splitting with high efficiencies. By studying reviews both old and new, inspiration can be drawn for future novel photocatalysts, improving efficiencies and creating new production methods. It is noted that the review will not focus on dye-sensitization, as this is out of the scope of the project due to long term feasibility issues associated with the stability of most well-known light absorbing dyes [67]. Furthermore, PEC systems, or coupled systems (photocatalyst + solar cell) will not be considered due to their complexity, and less feasible nature for this project which is to screen efficient materials [68].

1.3.1 UV-Active Semiconductors

In 1972, Honda and Fujishima showed that TiO_2 could be used to split water when irradiated with UV-light. They used a PEC cell, with Pt as the counter electrode, and an external bias. Hydrogen was evolved on the Pt electrode, and oxygen on the TiO_2 electrode [6]. After this experiment, extensive research began on water splitting, in an effort to generate a more efficient photocatalyst which could readily produce clean, renewable fuel. TiO_2, the quintessential photocatalyst, has been continually investigated through various methods, including doping with different elements, altering particle size and morphology, and using different cocatalysts. In the mid 80's, many groups began changing experimental conditions and adding in various cocatalysts to improve performance. Yamaguti and Sato found that by loading TiO_2 with Rh and using NaOH, an enormous increase in gas evolution rates could be achieved; with a QY of 29 % at 365 nm [69]. Selli et al. reported enhanced performance when a pH difference was employed between a TiO_2 electrode and Pt counter electrode [70]. At a similar time, Gratzel's group reported a series of experiments which focused on a TiO_2 colloid loaded with Pt and RuO_2 as a reduction and oxidation catalyst respectively. They found this further increased the quantum yield, and also increased the hydrogen evolution rate [71]. Hydrogen was found to be produced at 2.8 cm^3 h^{-1}, much higher compared to 1 cm^3 h^{-1} without RuO_2 [72]. Oxygen gas was not detected at the beginning of the experiment because it was assumed to strongly adhere to the catalyst in the initial period. Instead of using anatase TiO_2, Fu et al. studied bicrystalline titania ($TiO_2(B)$) [73]. They synthesised and investigated a heteropoly-blue sensitizer and Pt loaded $TiO_2(B)$ nanoribbon for water decomposition for hydrogen evolution (quantum yield (QY) = 8.11 %). Their result showed that a mixture of $TiO_2(B)$ and anatase gave the maximum quantum yield, suggesting that interfacial charge separation improved the efficiency. In another report by this group, the solvent effect of water splitting was investigated [74]. The hydrogen production rate was found to be increased in the presence of monochloroacetic acid and dichloroacetic acid with Pt/P25 as the photocatalyst. TiO_2 remains limited by its large band gap which prohibits light absorption in the visible range. Despite efforts to alter its characteristics, it still yields low efficiencies as dopants only offer a limited number of donor/acceptor states in comparison to the large number of states available at Ti3d and O2p orbitals [60]. Scientists have also experimented with other UV materials which share similarities with TiO_2 and explored their characteristics. Metal oxides which respond to UV light commonly have a full valence band which is mainly constructed of oxygen 2p electron orbitals, and a conduction band with empty metal d orbitals for transition metals and s or p orbitals for main group metals.

Domen's group paid particular attention to $SrTiO_3$—a perovskite structure material, part of the 'titanates' [75–79]. $SrTiO_3$ has a band gap of 3.2 eV, so responds to UV light ($\lambda < 387$ nm). They examined the use of NiO as a cocatalyst, and found that both oxygen and hydrogen can be evolved in a stoichiometric ratio—pure water splitting. The NiO cocatalyst was prepared by a H_2/O_2 redox reaction on NiO to form a Ni/NiO_x double layer. Kim et al. studied another important titanate, $La_2Ti_2O_7$. Lanthanum titanate is a layered structure consisting of four TiO_6 unit slabs separated by La^{3+} ionic layers [80]. When $La_2Ti_2O_7$ splits water, H_2 and O_2 can be produced in high amounts (QY up to 12 %) in the presence of a NiO cocatalyst. It can be even further enhanced by adding BaO as a dopant, and increasing pH with NaOH; the QY increases to 50 %.

Rutile zirconia (ZrO_2) has a very large band gap (5.0–5.7 V), but can split water without a cocatalyst under UV illumination [81]. Sayama and Arakawa reported the use of alkali carbonates in water; $NaHCO_3$ gave the highest efficiency in water splitting [82]. Interestingly, here, the addition of a cocatalyst actually decreased the ability of ZrO_2 to split water. This suggests that there is a large energetic barrier between the cocatalyst and semiconductor as a result of the large bandgap and position of the CB and VB. Jiang et al. turned their attention to ZrW_2O_8 in order to overcome the problem of ZrO_2 having large band gap [83]. ZrW_2O_8 was estimated to have a band gap of about 4.0 eV. It showed ability to decompose water to yield hydrogen or oxygen in the presence of an appropriate electron or hole scavenger under a 300 W Hg-Xe lamp. The production rate was not high, e.g. 23.4 μmol h^{-1} for H_2 and 9.8 μmol h^{-1} for O_2.

Tantalates and niobates have shown some of the highest efficiencies recorded for photocatalysis. A large number of reports focus on combining an alkaline earth metal with a tantalate: (M represents different alkaline metals; Na, K, Rb, etc.) $MTaO_3$ [84–86], MTa_2O [87], $M_2Ta_2O_7$ [88], and $M_2Nb_2O_7$ [89]. These materials are capable of producing both H_2 and O_{-2}, but only when illuminated with high energy UV light. The perovskite structure, as seen in $SrTiO_3$, is also present within both tantalates and niobates; the ABO_3 structure allows a wide range of options to be explored, as many different metals can fit into the A and/or B sites. It was reported that sodium tantalate ($NaTaO_3$) exhibits a particularly high quantum yield

(20 %, $\lambda = 270$ nm) for both hydrogen and oxygen production, when used in conjunction with a NiO cocatalyst [87]. In further work, it was found that doping the NaTaO$_3$/NiO compound with lanthanum (2 mol%) resulted in the quantum yield increasing to 56 %, the current record for UV driven water splitting. It is believed that the addition of La not only reduces the particle size (increasing the overall specific surface area), but also helps to separate reduction and oxidation sites.

M$_4$Nb$_6$O$_{17}$, layered niobate structures, were reported to be useful for overall water splitting [84]. It has been suggested that these structures enable the production of both H$_2$ and O$_2$ in two different layers. This would be an advantageous technique for splitting water—as separate evolution sites are known to prevent recombination, but whole separate layers would inhibit recombination even further. Although M$_4$Nb$_6$O$_{17}$ materials are able to split water without any cocatalyst at relatively low efficiencies, it was reported that hole scavengers can increase H$_2$ production dramatically [90].

Studies have also been carried out regarding the comparison of lattice parameters regarding niobate and tantalate perovskites [91]. The authors claim that phonon modes in semiconductors play a vital part in regulating the migration of charge carriers, and can be the deciding factor concerning performance. A relationship between Raman spectra results and the catalytic performance shows that a red shift in phonon frequency leads to a decrease in activity, essentially overriding influences of particle size and band gap width. Perovskites have the ability to accept many different metal dopants, and efforts are being made to enhance not just UV-active materials, but their visible light driven counterparts as well.

1.3.2 Semiconductors Activated by Visible Light

Metal oxides in the main group of the periodic table have also been subject to extensive research for the purpose of water splitting. Elements in groups 13 and 14; In, Sn, Sb, Ga and Ge have all been reported to be active for water splitting when combined with suitable cocatalysts. Sato et al. reported MIn$_2$O$_4$ (in this case M represents Ca, Sr and Ba) had distorted InO$_6$ octahedra in the lattice, showing H$_2$ and O$_2$ production from pure water when a cocatalyst (RuO$_2$) was added to the materials, through an impregnation method [92, 93]. CaIn$_2$O$_4$ exhibited the highest activity under UV light. Tang et al. examined different MSnO$_3$ (M = Ca, Sr and Ba) colloids, and obtained both hydrogen and oxygen in the presence of the RuO$_2$ cocatalyst [94]. SrSnO$_3$ has the highest photocatalytic activity among common tin oxides, due to the suitable band position and rapid charge transfer, mainly due to the distorted connection of SnO$_6$ in SrSnO$_3$, and also due to the strong electron-lattice interaction. The study of the SrSnO$_3$ morphology found that preparation method can lead to very different morphologies, which in turn alter performance. SrSnO$_3$ nanorods prepared by a hydrothermal method led to a tenfold increase in photocatalytic activity, in comparison to standard spherical particles (high temperature solid state reaction) which have similar surface area and optical density [95]. Sato et al. have also found the ability of $M_2Sb_2O_7$ structures (M = Ca, Sr), which also contain octahedral (SbO_6) connections, to split water into hydrogen and oxygen in near-stoichiometric ratio under UV radiation and with the addition of RuO_2 [96]. Gallium and germanium oxides have also been explored; water splitting is possible with Ga_2O_3 and Zn_2GeO_4. Interestingly, reports show that doping Ga_2O_3 with Zn can result in increases of up to 8 times in photocatalytic performance [97–99].

Metal sulphides and nitrides also have the ability to dissociate water. The valence bands are often sulphur 3p, or nitrogen 2p, which results in a contraction of band gap as S3p and N2p orbitals often lie at more negative potentials than that of O2p. ZnS has been one of the most extensively tested materials from the non-oxide group, even though its 3.6 eV band gap prohibits visible light absorption. Yanagida et al. reported hydrogen evolution using a relatively weak 125 W Hg lamp, and used the relatively uncommon hole scavenger, tetrahydrofuran [100]. Another report changed a great deal of variables, and methodically examined the photocatalytic activity of zinc sulphide; altering the hole scavenger, pH and temperature. A record 90 % quantum yield was attained for hydrogen production, at 313 nm with a platinum cocatalyst, immersed in a solution of Na_2S, H_3PO_3 and NaOH [101]. Fujishima's group evolved hydrogen from the semiconductors $CuInS_2$ and $CuIn_5S_8$. They were suspended in water with a sulphite hole scavenger under UV illumination [102]. However sulphides often undergo self-degradation via photocorrosion (typically photooxidation of sulphur ions), therefore it is vital that a sulphite/sulphate scavenger is used to replenish the degraded semiconductor.

A string of reports have been published on UV responsive nitride photocatalysts, investigating the feasibility and exploring the band structure [103–107]. The DOS reports suggest valence bands are made up of nitrogen 2p electron orbitals. This has the effect of narrowing the band gap. Ge_3N_4 was the first non-oxide, UV-active material for overall water splitting. Its VB is structured by N2p orbitals, and its CB is a hybrid of Ge4s4p. The reported quantum yield is 9 % at 300 nm when used in conjunction with a RuO_2 cocatalyst. Gallium Nitride (GaN) is already a widely used material in blue LED's, violet laser diodes, and high speed field effect transistors. It has a band gap of 3.4 eV, which enables it to only effectively use radiation with $\lambda < 365$ nm. Water splitting is possible using GaN (with a Rh-Cr cocatalyst complex), however, when combined with ZnO in a solid solution—another UV-active compound, the joint GaN:ZnO solid solution (achieved due to same wurtzite structure and similar lattice parameters) responds to visible light. Computational studies present strong evidence that this is due to p-d repulsion in electron orbitals [108], which can be found in O2p–Zn3d and it is hypothesised that this same phenomenon appears in N2p–Zn3d orbitals; essentially moving the valence band upwards to higher potentials—narrowing the band gap to around 2.4–2.8 eV. Even though nitrides and sulphides somehow suffer from photocorrosion, research is still active, since a lot can still be learnt through experimentation—evident for example from GaN:ZnO research.

1.3.3 Semiconductors Activated by Visible Light

To reiterate, an efficient photocatalytic system for water splitting is the target in the field, as visible light offers enough energy to meet industrial efficiencies. It is widely accepted a single photocatalyst for both H_2 and O_2 production from water is preferable due to its simplicity, however since that goal is considered by some to be extremely difficult to attain, significant developments have been made in designing photocatalysts that are active for half reactions—either hydrogen or oxygen. These photocatalysts, if active enough, can be used in Z-Schemes for overall water splitting, or to produce hydrogen by utilising chemical feedstocks and biomass in half reactions [109–111]. Oxygen produced from photocatalytic half reactions can also be used as a pure O_2 source for fuel cells. However, in comparison to UV active photocatalysts, the number of pure phase undoped visible light active photocatalysts are few in number, with even less showing reasonable activity for half reactions and good stability.

Usually, stable semiconductor photocatalyst metal oxides are composed of O2p orbitals and transition metal d^0 or d^{10} shells, with approximate band gaps around 3 eV due to the valence band position. Despite having the necessary redox potentials to prevent photocorrosion, UV-active photocatalysts are unable to provide the necessary efficiency required for industrial application. Therefore, one of the first methods researchers used to reduce the band gap of UV-active materials was to dope anions (e.g. N, C and P) or cations (e.g. Cr, Ni, Ta, Sb, Fe) to provide new valence and conduction band positions (Fig. 1.9).

Photocatalysts such as TiO_2 have been extensively doped since 1982, Borgarello et al. pioneered a pentavalent chromium dopant which enabled the production of both hydrogen and oxygen under visible light ($400 \leq \lambda \leq 500$ nm) [112]. Afterwards, various groups doped TiO_2 with metals such as V, Mn, Mo, Fe, Ni, Sn, etc., however, results show that quantum yields for any discrete visible light

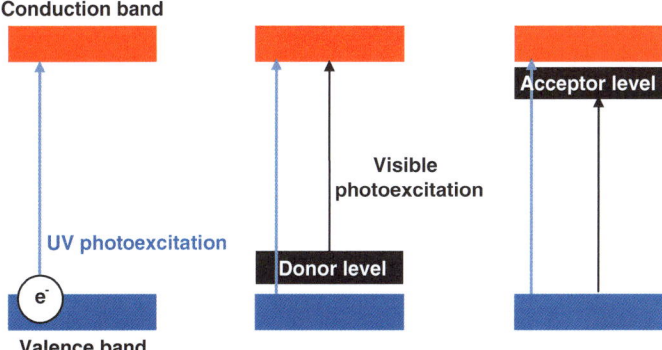

Fig. 1.9 Formation of new valence and conduction bands by electron donor and acceptor atoms, enabling a UV-active material to respond to visible light

wavelength range did not exceed 1 % [113–117]. In some cases, metal ion species act as both light harvesting centres (band gap change), and also as electron/hole traps, preventing recombination. This is especially evident when Ag^+ and Pt^{4+} were used and activity increased considerably [118, 119]. On the contrary, certain metal ions such as Co and Al have a detrimental effect on photocatalytic activity, presumably acting as recombination centres and preventing charge transfer to the surface [117]. The prototypical non-metal doping of titania was demonstrated by Asahi, who successfully increased the absorption spectrum to over 500 nm by doping with nitrogen [60]. Unfortunately, water splitting is not possible with $N-TiO_2$, as the kinetic requirements for oxidation are not satisfied by the potential of the N2p dopants. Organic degradation is however possible, and many reports have given evidence to the mechanisms behind the enhancement [42, 61]. Various other dopants such as sulphur, carbon, boron, and fluorine among others are able to be successfully incorporated into the crystal structure of TiO2 [120–122], however all displaying quantum efficiencies of less than 1 % at visible light wavelengths.

It wasn't until Kudo et al. began doping the perovskite $SrTiO_3$ with metals such as Cr, Ru, Ir, Mn, Sb, Ta, Ni, and Rh, that quantum efficiencies started to increase [123]. Rh (1 %) doped strontium titanate demonstrated a quantum efficiency of 5.2 % for hydrogen production at 420 nm, the highest value among the different dopants. Rhodium forms an electron donor level more negative than that of oxygen 2p orbitals, narrowing the light absorption into the visible range. However, oxygen production was nearly inhibited, presumably due to Rh_xO_y surface species blocking certain active sites. The unique redox functionality of $Rh^{3+/4+}$ is thought to play a part in the enhanced activity. Rhodium can also be doped into other perovskites such as $CaTiO_3$, and $Ca_3Ti_2O_7$, however due to the complex band structure, the Rh dopant was indifferent, and efficiencies were immeasurably small [124, 125].

Wolframite tantalates also showed activity when doped with various metals. Zou et al. reported that nickel doped $InTaO_4$ can split water in stoichiometric amounts without any electron donor/acceptor with a QY of 0.66 % (402 nm) [126]. However, it must be in the presence of a RuO_2 or NiO cocatalyst. Doped $NiO-InTaO_4:Ni$ has also been investigated and found to be less active due to having a larger band gap (2.6 eV undoped, 2.3 eV doped). This is because the In^{2+} ions are dislodged by Ni^{2+}, which consequently causes the lattice to contract, and in turn reduces the band gap by the introduction of partially filled nickel 3d orbitals. $NaBO_3$, a series of known UV-active materials, were investigated by Kudo et al. (B = Nb or Ta) [127]. $NaNbO_3$ doped with iridium and either Sr, Ba, or La shows activity for separate hydrogen and oxygen evolution. On the other hand $NaTaO_3$ doped with iridium and Sr/Ba/La only showed activity for hydrogen evolution due to the change in band structure and blocking of oxygen evolution sites.

Due to the nature of a dopant, i.e. the incorporation into the lattice of a foreign element, the number of electronic states is inherently low. Therefore, upon excitation, the new band in the otherwise forbidden region only can provide a limited amount of photoexcited electrons to the conduction band for reduction reactions. The same amount of limited holes are left behind for potential oxidation. In other words, the contribution to the density of states by this dopant is small in comparison to that of the bulk band structure, and therefore the maximum efficiency will also be low. Furthermore, one cannot simply increase dopant amount to accordingly increase states, otherwise crystal structure will slowly change/buckle, and original band structure will deteriorate. Ideally therefore, logic would dictate to focus research on the discovery of new pure phase materials, or by enhancing suitable pre-existing compounds.

One group used an ingenious way of circumventing crystal structure deformation by creating a solid solution, whereby two wide band gap semiconductors create one narrow gap solid solution; GaN:ZnO (Fig. 1.10). Domen et al. produced a range of visible light nitrides/oxynitrides [128–131]. The nitrides/oxynitrides need scavenger species to function, and also need the pH to be adjusted by a lanthanum containing material. Systems comprising of solid solutions of oxides and nitrides have been reported with some positive results [18, 132, 133]. GaN:ZnO ($Ga_{0.88}$ $N_{0.88}Zn_{0.12}O_{0.12}$), is able to split water with a high QY under visible illumination without scavenger species (QY = 5.9 %, 420 nm). In order to obtain this record high QY for overall water splitting, the pH must be adjusted to 4.5 (using H_2SO_4), and the gallium-zinc composite must be coated with $Rh_{2-x}Cr_xO_3$ cocatalyst nanoparticles (10–30 nm diameter). If the cocatalyst isn't present, no water splitting is observed. Both GaN and ZnO have an almost identical crystal structure, including space group and lattice parameters, and therefore the crystal structure is not deformed when a solid solution is formed.

In terms of native oxide photocatalysts, WO_3 is well-documented for oxygen production in water in the presence of an electron scavenger such as Ag^+; however the low conduction band prevents hydrogen production. Attempts of incorporating different elements into tungsten oxide have been successful, and hydrogen production is a possibility by creating new LUMO and HOMO levels. $AgInW_2O_8$,

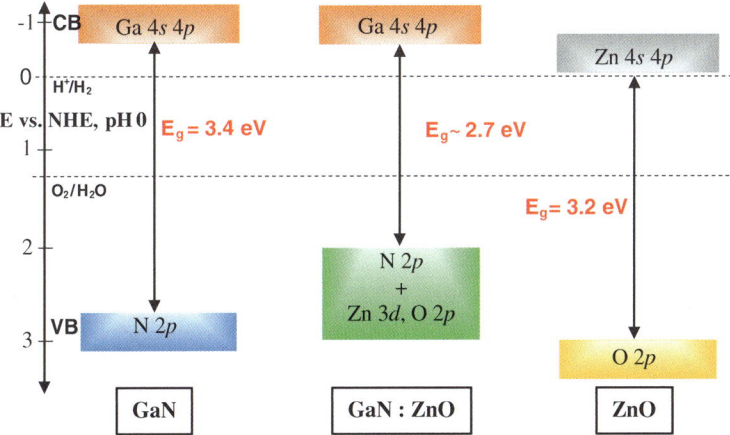

Fig. 1.10 Illustration of GaN:ZnO solid solution. Conduction band orbitals remain as Ga 4 s 4p, whilst valence band is hybridised to N2p, Zn3d and O2p. Adapted from Maeda et al. [132]

$AgBiW_2O_8$ and Ag_2WO_4 have been experimentally proven to produce both O_2 and H_{2-}, albeit in relatively low yields [134, 135].

A large amount of reports have been published in the last 15 years or so on $BiVO_4$ (monoclinic crystal structure), some with quantum yields of up to 9 % for oxygen production under visible light (450 nm) irradiation [17, 31, 136]. It was found that the activity of $BiVO_4$ is heavily dependent on the preparation method and crystal phase—in particular, the monoclinic phase is principally the most active. The efficiency has also been shown to be irrespective of surface area and more influenced by defect concentration; high temperature solid state reactions yield low efficiencies, and wet chemical methods promote higher yields due to phase purity. $BiVO_4$ cannot in fact reduce water to produce hydrogen due to its low conduction band (thought to be between 0.1 and 0.7 eV vs. NHE). Another monoclinic material, $In_xMTi_yO_z$ (M = Ni, Cr, or both) was investigated, yielding results which suggest that hybridised electron orbitals lead to a higher charge carrier transportation.

Sulphides have potential as visible light photocatalysts for the production of solar hydrogen; but their application has been limited due to photocorrosion. CdS has been stringently investigated due to its narrow band gap and ideal band positions for both O_2 and H_2 evolution. Two groups studied CdS, reporting hydrogen production using edetic acid as an electron donor [137, 138]. They found that with the addition of Pt cocatalyst, H_2 evolution rates will increase by a factor of 10. By introducing Na_2SO_3 into the solution a quantum yield of 35 % (at 430 nm) can be achieved. Progressing from this, if CdS is loaded with a cocatalyst comprising of Pt-PdS, a quantum yield of 93 % can be achieved for photocatalytic hydrogen production [139]. Unique to sulphide photocatalysts, another sulphide (Rh_2S_3, Ru_2S_3, MoS_2) can be added to CdS or $CuGa_3S_5$ as catalysts, and perform better than their metallic counterparts [140–142]. The activity increase is attributed to a better coupling between structures, i.e. the presence of sulphur in the catalyst enables a strong metal support interaction between substrate and metal. This in turn increases electron-hole separation and consequently more probable reduction reactions. In an alternate report, Gratzel's group have shown that loading CdS with Pt and RuO_2 leads to a decrease in photocorrosion [143]. In order to increase the resistance to photocorrosion, electron donors such as S^{2-}, SO_3^{2-} must be introduced into the solution. Various strategies have been employed to overcome such problems as; encapsulation of CdS nanoparticles [144], formation of CdS composites with TiO_2 [145] and ZnS [146], and doping of CdS by noble metals and transition metals [147] but with limited success.

A recent report from Ye et al. have found that a phosphate compound could function as a photocatalyst with a record-breaking quantum yield [7]. Silver orthophosphate (Ag_3PO_4) has a maximum quantum yield for O_2 evolution of 89 % at 420 nm, yet is only 50 % at 500 nm. However, the orthophosphate compound does require the use of an electron scavenger to inhibit self-decomposition. The band gap was measured by UV-Vis to be approximately 2.4 eV. The conduction band is more positive than 0 V versus NHE, and thus the compound cannot produce hydrogen from water. The valence bands are made up of hybridised silver 4d and oxygen 2p orbitals, while the conduction band is composed mainly of hybridised silver 5s5p, and a small amount of phosphorus 3 s electron orbitals [7]. These highly dispersive conduction bands are thought to provide an efficient charge transfer system with low charge recombination, and result in the compound's high photooxidation. A computational study by Umezawa et al. used DFT to ascertain the exact mechanisms behind the remarkable activity [148]. They found that the formation of strong P-O bonds in PO_4 tetrahedra deteriorates the covalent nature of Ag-O bonds. This decreases the hybridisation between Ag d and O p bands, and as a consequence, excludes the d character from the conduction band minimum. This then leaves dual Ag s—Ag s hybrid states, with the absence of localised d character. As an interesting side point, the DFT results highlight the isotropic nature of the electrons in Ag_3PO_4; the mass of electrons (m_e^*) is small in all directions. This is the opposite of holes, whereby the mass is anisotropic in three different primary directions; <100>, <110>, and <111> as later reported.

Currently, no efforts have been made in order to stabilise the compound, and little follow up literature on water photooxidation is available. Bi et al. did however test different exposing facets for the photodegradation of organic dyes (methyl orange—'MO', methylene blue—'MB', and Rhodamine B—'RhB'), for which there was a difference in degradation rate depending on what facet was predominant ({100}—cubic, or {110}—rhombododecahedron) [149, 150]. According to the reports, the two facets tested showed considerably higher activity for organic degradation than $N-TiO_2$ and multifaceted (roughly spherical, and therefore a mixture of all facets) Ag_3PO_4. The group continued to mention that although the enhanced activity did not correlate with surface area, that is a higher surface area did not give a higher activity, a connection between the DFT calculated surface energy and activity is evident. As the surface energy increases, so too does the degradation rate of organic contaminants. A higher energy surface facet (such as the low index {100} or {110} planes) is inherently more 'unstable', and therefore possesses a rougher or more 'kinked' surface. From calculations of Ag_3PO_4, it can be concluded that a higher number of oxygen vacancies create more active sites on the surface to which the organic molecule can adsorb—thereby enabling the degradation reaction to occur faster. However since the energetic requirements of organic degradation are different than that of water splitting, it was not clear if the 'facet effect' would be beneficial to water splitting in this particular case.

In 2008, an organic-polymer-based photocatalyst, graphitic carbon nitride ($g-C_3N_4$), was found to show sufficient redox power to dissociate water under visible light in a suspension system [151]. The compound was fabricated by simply calcining cyanamide in either air, or a nitrogen atmosphere. Historically, the structure of carbon nitride has been debated since initial reports in 1834; with melon, melem and melam structures all being proposed [152–154]. It is now accepted that the graphitic form of carbon nitride, which is photocatalytically active, is initially composed of stable C-N heptazine (tri-s-triazine) sheets, hydrogen bonded in a zig-zag formation [155]. Upon further condensation and removal of NH_3 groups, heptazine units continue to polymerise to form what is now conventionally known as graphitic carbon nitride (Fig. 1.11) [156, 157]. The first experimental example of $g-C_3N_4$ showed that the compound could evolve hydrogen with and without a cocatalyst. However, the presence of photodeposited platinum increased yields

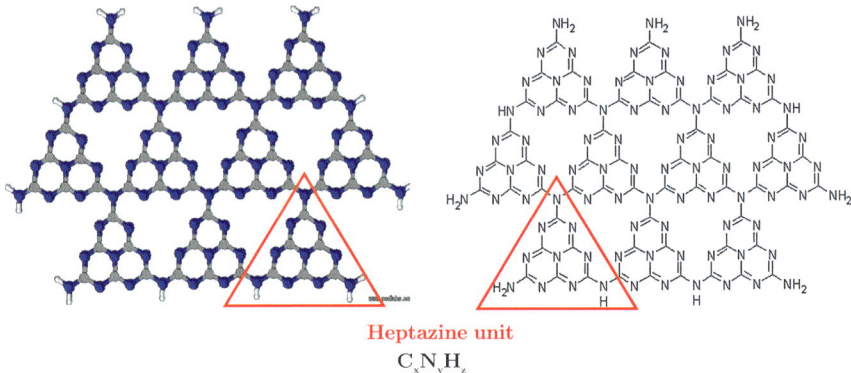

Heptazine unit

$C_x N_y H_z$

Fig. 1.11 Ball and stick (*left*), and molecular schematic drawing of g-C$_3$N$_4$ (*right*), using Chem-Sketch©. A single heptazine unit is highlighted in *red*. Graphitic carbon nitride is composed of; carbon (*grey*), nitrogen (*blue*), and hydrogen atoms (*white*)

more than 2.5 times. DFT calculations confirmed UV-vis results that the band gap was approximately 2.7 eV. Further calculations revealed that the conduction band is made up of carbon p$_z$ orbitals, whilst the valence band is primarily composed of nitrogen p$_z$ orbitals. Both LUMO and HOMO are contained within the heptazine monomer and no charge is transferred between heptazine units. Tertiary nitrogen merely links heptazine units together, however no reports to date have shown an isolated heptazine unit to itself be a functional photocatalyst—meaning that the role of linking nitrogen atoms could be crucial to the activity of g-C$_3$N$_4$.

Due to the stability and elemental abundance, intensive research efforts have been made in order to potentially increase the efficiency. Research efforts by Dong et al. have shown that polymeric carbon nitride can be fabricated not just from cyanamide, but derivatives such as dicyandiamide (DCDA), thiourea, and urea [158, 159]. The latest documented quantum yields under visible light for H$_2$ production from water using g-C$_3$N$_4$ however, do not exceed 4 % (excluding dye-sensitized systems) [151, 160–166], which is far from the requirements demanded by researchers [66]. Researchers have also made modifications to g-C$_3$N$_4$ based photocatalysts, e.g. via nanocasting [165], soft-templating [167] and chemical functionalization, i.e. by post [168] or in situ heteroatom doping [167]. Despite the various methods used to increase efficiency, the highest quantum efficiency remains at 3.75 % for hydrogen evoltuion under visible light (420 nm) [165].

The first of two ways to achieve colloidal overall water splitting under visible light is to use a single photocatalyst; the standout (and only working example) is GaN:ZnO (with Rh-Cr catalyst), achieving a quantum yield of 5.9 % at a working wavelength range of 420–440 nm [132]. The other is nickel doped InTaO$_4$ (NiO$_x$ catalyst), which has demonstrated a quantum yield of 0.66 % at 402 nm [126]. The main issue with the two aforementioned systems lies with the exclusivity, that is, it is difficult to systematically build upon the existing research because of the special-ised nature of the compounds. For example, gallium nitride and zinc oxide are an almost perfect match with one another in terms of crystal structure. That coupled with the fortuitous band structure yields a photocatalyst system which is unique.

The second approach to visible light water splitting is to use a two photon process, termed a 'Z-Scheme'. Inspired by natural photosynthesis, Bard proposed a biomimetic analogy composed of two inorganic semiconductor photocatalysts in 1979 and recently there have been some successes in the field based on various different photocatalysts [169, 170].

1.3.4 Z-Scheme Systems

Nature splits water into O_2 and the equivalent H_2 species by a double excitation process in PSII ('Photo-System') and PSI, instead of a single excitation, in which the two half reactions are spatially separated and take place in PSII and PSI, Fig. 1.12. This overcomes the main problems of a singular photocatalytic water splitting system both kinetically and thermodynamically, as well described in a previous review [171]. In short the Z-scheme has the potential to be more practical than a single photocatalyst [172].

Briefly, a Z-scheme system is a combination of two semiconductor photocatalysts, in which each photocatalyst is responsible for one half reaction and a soluble mediator helps electron transfer between the two photocatalysts so that an ideal cycle can be completed. The mediator is very important because it dramatically inhibits the fast unfavourable recombination of charge, analogous to the electron transport chain between PSII and PSI. Given such advantages of a double excitation process, there are many researchers working on either two photocatalysts each of which favours either H_2 or O_2 production, or a new mediator to efficiently transfer charge between two photocatalysts.

Initially, Darwent and Mills showed that the reduction potential of Fe^{3+} (from $FeCl_3$) was near the conduction band of WO_3, i.e. the iron ion acted as an electron scavenger and thus WO_3 could produce oxygen under visible light from water [173]. This observation was also confirmed by Graetzel [12]. In 2001, Sayama collaborated with Abe and created a UV-responsive anatase TiO_2—rutile TiO_2 dual photocatalyst z-scheme system [174], using an iodide (redox couple: I^-/IO_3^-, from NaI) mediator inspired by the early works of Kim et al. [175]. This mediator was also used in a visible light system based on Pt-loaded mixed Cr/Ta doped $SrTiO_3$, and PtO_x–WO_3, which can achieve a quantum yield of 1 % for overall water splitting [176]. In a water splitting reaction, the redox couples behave as follows:

$Fe^{2+/3+}$ redox potentials

$$O_2\ photocatalyst : \ Fe^{3+} + e^- \rightarrow Fe^{2+}$$
$$2H_2O + 4h^+ \rightarrow O_2 + 4H^+$$
$$H_2\ photocatalyst : \ Fe^{2+} + h^+ \rightarrow Fe^{3+} \tag{1.29}$$
$$4H^+ + 4e^- \rightarrow 2H_2$$

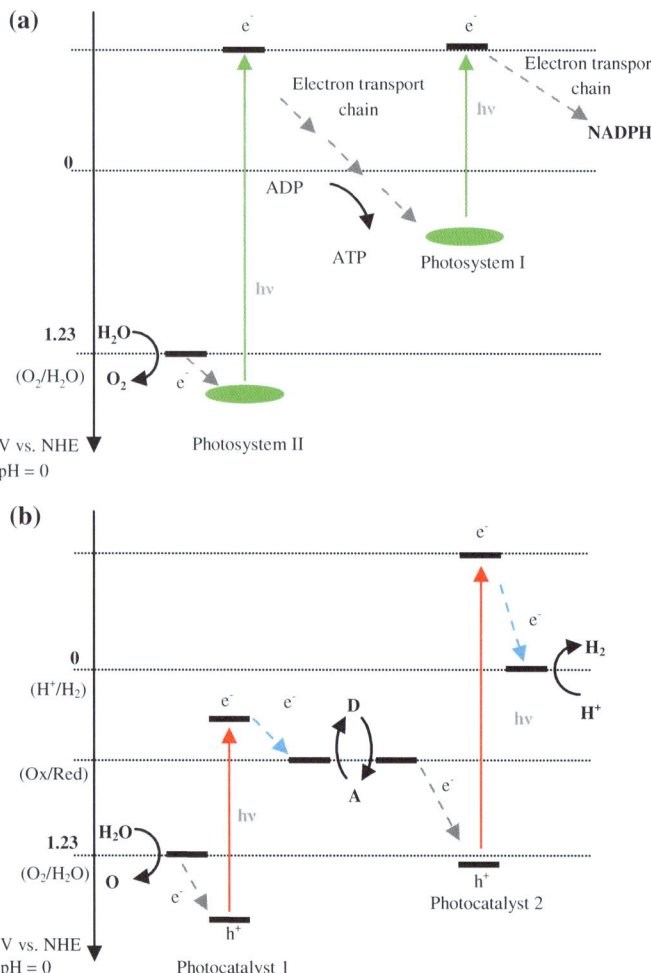

Fig. 1.12 a Natural photosynthesis. **b** An artificial analogy involving two different photocatalysts possessing two different band positions (A is electron acceptor, D is electron donor) [172]

I^-/IO_3^- redox potentials

$$H_2 \, photocatalyst : I^- + 3H_2O + 6h^+ \rightarrow IO_3^- + 6H^+ \, (0.71 \, V \, vs. \, NHE)$$
$$6H^+ + 6e^- \rightarrow 3H_2$$
$$O_2 \, photocatalyst : IO_3^- + 6e^- + 3H_2O \rightarrow I^- + 6OH^-$$
$$2H_2O + 4h^+ \rightarrow O_2 + 4H^+$$

$$(1.30)$$

The activity of rutile TiO_2 and WO_3, according to reports, stems from their suitable surface and band structure for the reduction of Fe^{3+} and IO_3^-, and more importantly, the selective oxidation of water, even in the presence of electron donors such as Fe^{2+} and I^- [177, 178]. Therefore, these are considered optimum water oxidation photocatalysts in Z-Schemes. In 2004, $BiVO_4$ was also shown to function as an oxygen evolution photocatalyst in the presence of the $Fe^{2+/3+}$ mediator, when paired with Pt loaded $SrTiO_3$ doped with Rh [170]. As mentioned previously $BiVO_4$ has a bandgap of approximately 2.4 eV, and therefore can absorb visible light up to 520 nm in most cases. In theory, a semiconductor with a smaller band gap can absorb more light, and potentially achieve higher efficiencies. However, researchers have shown that even by using an O_2 photocatalyst such as $BiVO_4$, which has a smaller band gap than either Bi_2MoO_6 or WO_3, the system is still limited in terms of efficiency by a H_2 photocatalyst $Pt/SrTiO_3$:Rh. Therefore, it would be advantageous to find a more efficient hydrogen evolving photocatalyst for use in a Z-scheme water splitting system. Further to this, another drawback of the 'Z-scheme' process are notably back reactions, and side reactions, both of which hinder efficiencies (the most common of which is where I^- is oxidised to I_3^-, which then cannot be reduced easily as it isn't an efficient electron acceptor.).

Various other photocatalysts can be used as a hydrogen evolution photocatalyst in a Z-scheme water splitting system, most notably nitrides and oxynitrides [179], as seen in Table 1.1. Tantalum oxynitride (TaON) has been shown to oxidise I^- to IO_3^- effectively, and when coupled with either WO_3 or Ta_2N_5, can participate in a Z-scheme system, with quantum yields up to 6.3 % at 420 nm—the highest reported to date [179, 180]. The authors claim that despite the high efficiency, the rate limiting step originally lies with the hydrogen production photocatalyst (TaON)—to which they added a protection layer of ZrO_2, helping to increase the thermal and chemical stability. As TaON is fabricated by nitridation of Ta_2O_5, and thus is an "intermediate phase" between Ta_2O_5 and Ta_3N_5, the crystallinity of the material is poor and results in many surface defects [181]. The authors then modified TaON with ZrO_2 in situ to maintain the Ta^{5+} cation at the surface (make Ta more 'cationic'), and to prevent the formation of lower valences, such as Ta^{3+} or Ta^{4+} which act as electron recombination centres and hinder efficiency. Despite tantalum's high cost, the system illustrates that nitrides and oxynitrides can be used in Z-scheme water splitting systems, and more importantly, due to the positioning of the redox potentials of both I^-/IO_3^- and Fe^{2+}/Fe^{3+}, be stable in both reduction and oxidation.

To briefly conclude, the latest studies show that quantum yields are relatively low for overall water splitting in the presence of a redox mediator, are no larger than 6 % under visible light, as seen in Table 1.1. However, many photocatalysts can potentially be used, as there is some flexibility with both band position and electron mediator. This bodes very well for future research, as the process is not as specialised as the aforementioned single photocatalyst water splitting system.

Table 1.1 Summary of water splitting systems using a Z-scheme setup, currently reported in the literature [181]

Entry	Photocatalyst (available wavelength)		Redox mediator	Maximum apparent quantum yield (%)	References
	H_2	O_2			
1	Fe^{2+}(<280 nm)	RuO_2/WO_3(<460 nm)	Fe^{3+}/Fe^{2+}	No available information	[182]
2	Pt/TiO_2 anatase (<380 nm)	TiO_2 rutile (<410 nm)	IO_3^-/I^-	1 (350 nm)	[174]
3	$Pt/SrTiO_3$:Cr/Ta(<550 nm)	PtO_x/WO_3(<460 nm)	IO_3^-/I^-	1 (420 nm)	[176]
4	$Pt/SrTiO_3$:Bi/Ga(<370 nm)	$SrTiO_3$:In/V(<370 nm)	IO_3^-/I^-	No available information	[183]
5	$Pt/SrTiO_3$:Rh(<520 nm)	$Ru/SrTiO_3$:In/V(<520 nm)	IO_3^-/I^-	0.33 (360 nm), 0.056 (420 nm), 0.039 (480 nm)	[184]
6	Pt/TiO_2 anatase (<400 nm)	NO_3^-(<250 nm)	NO_3^-/NO_2^-	no data	[185]
7	$Pt/SrTiO_3$:Rh(<520 nm)	$BiVO_4$(<520 nm)	Fe^{3+}/Fe^{2+}	0.4 (420 nm), 0.3 (440 nm)	[170]
8	$Pt/SrTiO_3$:Rh(<520 nm)	Bi_2MoO_6(<450 nm)	Fe^{3+}/Fe^{2+}	0.2 (440 nm)	[170]
9	$Pt/SrTiO_3$:Rh(<520 nm)	WO_3(<460 nm)	Fe^{3+}/Fe^{2+}	0.5 (420 nm), 0.2 (440 nm)	[170]
10	$Ru/SrTiO_3$:Rh(<520 nm)	$BiVO_4$(<520 nm)	Fe^{3+}/Fe^{2+}	0.3 (420 nm)	[186]
11	$Ru/SrTiO_3$:Rh(<520 nm)	$BiVO_4$(<520 nm)	none	1.7 (420 nm)	[187]
12	$Ru/SrTiO_3$:Rh(<520 nm)	$BiVO_4$(<520 nm)	Solid—reduced graphene oxide	1.03 (420 nm)	[188]
13	Pt/TaON(<520 nm)	PtO_x/WO_3(<460 nm)	IO_3^-/I^-	0.5 (420 nm)	[189, 190]
14	Pt/TaON(<520 nm)	$RuO_2/TaON$(<520 nm)	IO_3^-/I^-	0.1 (420–440 nm)	[191]
15	$Pt/ZrO_2/TaON$(<520 nm)	PtO_x/WO_3(<460 nm)	IO_3^-/I^-	6.3 (420 nm)	[180, 192]
16	$Pt/ZrO_2/TaON$(<520 nm)	$Ir/TiO_2/Ta_3N_5$(<600 nm)	IO_3^-/I^-	No available information	[179]
17	$Pt/ZrO_2/TaON$(<520 nm)	$RuO_2/TaON$(<520 nm)	IO_3^-/I^-	No available information	[193]

1.3.5 Effect of Morphology, Crystallinity and Size on the Activity of Photocatalysts

In Sects. 1.3.1 and 1.3.2 it was noted that both band structure and magnitude could contribute to the performance of a photocatalyst, however, this is one of just many factors which come into play. Cocatalyst choice, surface area, number of defects, among others all play vital parts in determining efficiency. An important point which was touched on in Sect. 1.1.4 was the morphology of particles. The three step process mentioned in Fig. 1.5 illustrates the importance of surface area and crystallinity. It is noted that the increase in crystallinity leads to a decrease in defects—which in turn decreases the recombination probability of charge carriers in the bulk and on the surface. On the other hand, high temperature calcinations which often result in good crystallinity, suffer from low surface area, and thus low number of surface active sites.

If a particle is of small size, on the nanoscale (10^{-9} m), then the distance the charge carrier (electron or hole) has to travel to reach the surface is decreased. However, decreasing a particles' size leads to a quantum sizing effect, effectively increasing the band gap [194, 195]. This arises when charge carriers are constrained to a specific portion of space by a potential barrier; the region is less than or equal to their de Broglie wavelength, and the energy states associated with the particle become quantized, as opposed to continuous. It can be difficult to achieve particles with high crystallinity, and correct size. Several methods are available which are used to fabricate highly crystalline nanoscale materials, including hydrothermal synthesis (both microwave assisted and conventional) and wet chemical techniques (e.g. sol-gel method, phase transfer, ion exchange). These techniques are often deployed depending on what temperature the reaction between two or more reactants happen. Dropping below room temperature for a certain reaction would yield smaller particles and reduce agglomeration/grain growth, however, for another; there would not be enough thermal energy to facilitate a reaction. Conversely, if a reaction happens at temperature T and pressure P, then exceeding these parameters will only result in the accumulation of particles. Microwave assisted hydrothermal synthesis is one of the most prominent methods of nanocrystal synthesis. Its advantages over conventional heating include; direct control over temperature and pressure within a sealed vessel, activation of rotational and vibrational nodes within molecules, high interaction with polar molecules and uniform heating. These all lead to an overall quicker reaction time, higher product yield and the use of much lower temperatures [196, 197]. Conventional hydrothermal synthesis involves simply heating a vessel in an oven or oil bath; the vessel is sealed and high pressures build up inside. Wang et al. found that by varying temperature they were able to selectively produce different sized nanocrystals [198]. Higher temperatures yield larger particles; Ag samples reacted at 90 °C resulted in 6 nm particles, whilst at a temperature of 180 °C, particles were found to be around 12 nm. This trend was observed for over 40 different materials; metals, semiconductors, dielectrics and ferrites.

Sol-gel synthesis falls under wet chemical techniques, and is widely employed in the field of nanoparticle preparation. Precursors develop from a simple solution, to a more complex network of discrete particles or chains, with the addition of a morphology controlling agent (sometimes termed 'surfactant') in a thermally controlled environment. A limitation of the sol-gel method is the activation (or burning off) of the surfactant—which can often increase particle size at high temperatures. This must be done since the agent is simply a cast for the particles to form in, and should not be included in experiments—especially in water splitting since some are organic and will act as electron donors. Parida et al. demonstrated that at an activation temperature of 500 °C, $LaFeO_3$ exhibited a 50 % increase in H_2/O_2 yield when compared with samples which were treated at 900 °C. particles of the lower temperature samples were also 4 times smaller and had a surface area which was over 5 times larger [32].

It is important to take into account the amount of cocatalyst on the surface; reports in which cocatalysts are used often vary dramatically in terms of loading (in wt% or mol%), but commonly reach a peak, depending on the type of photocatalyst and particle size [139]. Too little cocatalyst results in incomplete coverage, whilst too much cocatalyst results in the surface of the photocatalyst being covered, and unable to absorb enough light to facilitate redox reactions.

Surface area can also play an important factor in determining the activity of a photocatalyst, with contradicting theories. This is evident in reports; Kudo et al. state that their large $BiVO_4$ particles with a small surface area have higher O_2 evolution rates than smaller particles with a larger surface area [31]. Contradicting this, Ye et al. find that their $NaNbO_3$ samples experience a threefold increase in O_2 production and sixfold hydrogen increase when the surface area of the samples are increased [139]. Another group reported a 72 fold increase in hydrogen production on nanocrystalline $La_2Ti_2O_7$ when the surface area was increased from 1 to $52 \ m^2g^{-1}$ using a hydrothermal method as opposed to the conventional solid state reaction [199]. It is important to get a balance between particle size and photocatalytic activity, however, it is even more important in photocatalysis that there is a good degree of crystallinity, as this will decrease recombination between charge carriers. Conversely, a high surface area would be advantageous in degrading organic complexes since surface adsorption is the dominating factor.

In Sect. 1.3.2, it was stated that different facets of Ag_3PO_4 showed varying activity for the degradation of organic contaminants, and this correlated with the surface energy of the facet. In fact, this is not the first example of facet enhanced photocatalysis. Normally, conventionally reactive facets are diminished during the crystal growth process due to surface energy minimisation. However, manipulating exposing facets on surfaces is a technique which has been investigated in sensing [200], electronics [201], magnetic memory devices [202], and catalysis [203]—with catalysis being the most relevant to semiconductor photocatalysis. Therefore under the correct conditions, the exposing facets can be controlled on metallic and metal oxide surfaces, often using a surfactant (morphology controlling agent). For example, anatase TiO_2 crystals are often composed of thermodynamically stable {101} facets, with an average surface energy of $0.44 \ J \ m^{-2}$. Very recently, led by theoretical predictions, {001} facets on TiO_2 were successfully prepared by using fluorine ions as a capping agent. The {001} facets possess a much higher catalytic activity under UV light due to higher surface energy, calculated to be 0.90 J m^{-2} (over 2 times higher than {101}) [204]. This finding was supported by many other groups who demonstrated that high energy, low index facets of TiO_2, are more reactive for both hydrogen evolution and organic degradation reactions [205, 206].

In order to attain specifically terminated facets, morphology controlling agents (MCAs), such as organic surfactants (e.g. PVP), or capping agents (e.g. fluorine ions) are commonly employed. Due to strong interaction with the substrate, complete removal of these MCAs is very challenging, but absolutely essential for effective catalytic activity in water splitting. One of the major research obstacles facing researchers today is developing a facile, MCA free synthesis route to attain clean, reactive facets, for example, by kinetic control [207]. In order to remove some surfactants, it is possible to simply prepare a suitable calcination process—however, there is a risk that the high energy facets will diminish due to surface reconstruction aided by an increased temperature.

Apart from Ag_3PO_4, faceted $BiVO_4$ crystals have also been fabricated for the photooxidation of water under visible light. Xi et al. reported the synthesis of monoclinic bismuth vanadate (m-$BiVO_4$) nanoplates with exposed {001} facets, in the absence of traditional surfactants [208]. Using ethanolamine and water as an azeotropic solvent, a bismuth vanadate precursor was made, and then crystallised in hydrothermal reactor at a relatively low temperature (160 °C /12 h). {001} faceted m-$BiVO_4$ demonstrated superior photodegradation of RhB, and also for the photooxidation of water. The authors state that the superior activity could stem from an increase in surface energy, or, an increase in hydrophobicity as a result of a low index exposing facet due to a different surface structure.

1.3.6 Conclusions

It is evident from the current review that in field of photocatalytic water splitting, not one system can demonstrate efficiency, cost-effectiveness and robustness on a scale which would prove economically viable (10 % minimum solar-to-hydrogen conversion efficiency). However, it is agreed in the field that a photocatalyst which responds to visible light is a must; ultraviolet radiation does not make up enough of the electromagnetic (EM) spectrum incident on the Earth's surface. By studying UV-active materials, of which some have excellent charge separation characteristics, and modifying them, the chance of attaining the set goal will dramatically increase.

Furthermore, it is evident that finding a 'perfect' photocatalyst which can split water into its constituent parts under visible light would be unfeasible at this present moment. One of the more promising areas of visible light water splitting is the development of photocatalysts which participate in a half reaction, and can be coupled to either a PEC/solar cell, or in a colloidal Z-Scheme. Therefore, based on the review of the current efficient photocatalysts of the present day, it appears the most suitable strategy would be to develop highly efficient, separate, photocatalysts for oxygen production and hydrogen production, exactly mimicking nature photosynthesis system, then in turn use high efficiency photocatalysts in order to create new Z-Scheme systems. This strategy has been applied in the thesis.

References

1. Oberthür, S. & Ott, H. E. (1999). *The Kyoto protocol: International climate policy for the 21st century*. New York: Springer.
2. Woodhouse, M., & Parkinson, B. A. (2009). Combinatorial approaches for the identification and optimization of oxide semiconductors for efficient solar photoelectrolysis. *Chemical Society Reviews, 38*, 197–210.
3. Hoffmann, M. R., Martin, S. T., Choi, W. Y., & Bahnemann, D. W. (1995). Environmental applications of semiconductor photocatalysis. *Chemical Reviews, 95*, 69–96.
4. Sigfusson, T. I. (2007). Pathways to hydrogen as an energy carrier. *Philosophical Transactions of the Royal Society A: Mathematical, Physical and Engineering Sciences, 365*, 1025–1042.
5. Somorjai, G. A., & Turner, J. E. (1984). Catalyzed photodissociation of water—The first step in inorganic photosynthesis? *Naturwissenschaften, 71*, 575–577.
6. Fujishima, A., Honda, K. & Kohayakawa, K. (1972). Electrochemical photolysis of water at a semiconductor electrode. *Nature, 238*.
7. Yi, Z., et al. (2010). An orthophosphate semiconductor with photooxidation properties under visible-light irradiation. *Nature Materials, 9*, 559–564.
8. Sayama, K., et al. (2006). Photoelectrochemical decomposition of water into H_2 and O_2 on porous $BiVO_4$ thin-film electrodes under visible light and significant effect of Ag ion treatment. *The Journal of Physical Chemistry B, 110*, 11352–11360.
9. Maeda, K., et al. (2010). Photocatalytic overall water splitting promoted by two different cocatalysts for hydrogen and oxygen evolution under visible light. *Angewandte Chemie-International Edition, 49*, 4096–4099.
10. Bard, A. J., & Fox, M. A. (1995). Artificial photosynthesis: Solar splitting of water to hydrogen and oxygen. *Accounts of Chemical Research, 28*, 141–145.
11. Serpone, N., & Pelizzetti, E. (1989). *Photocatalysis: Fundamentals and applications*. Chichester: Wiley.
12. Erbs, W., Desilvestro, J., Borgarello, E., & Graetzel, M. (1984). Visible-light-induced oxygen generation from aqueous dispersions of tungsten (VI) oxide. *The Journal of Physical Chemistry, 88*, 4001–4006.
13. Nozik, A. J., & Memming, R. (1996). Physical chemistry of Semiconductor-Liquid Interfaces. *The Journal of Physical Chemistry, 100*, 13061–13078.
14. Reber, J. F., & Rusek, M. (1986). Photochemical hydrogen production with platinized suspensions of cadmium sulfide and cadmium zinc sulfide modified by silver sulfide. *The Journal of Physical Chemistry, 90*, 824–834.
15. Murphy, A. B., et al. (2006). Efficiency of solar water splitting using semiconductor electrodes. *International Journal of Hydrogen Energy, 31*, 1999–2017.
16. Gerischer, H. (1975). Electrochemical photo and solar cells principles and some experiments. *Journal of Electroanalytical Chemistry and Interfacial Electrochemistry, 58*, 263–274.
17. Kudo, A., Omori, K., & Kato, H. (1999). A novel aqueous process for preparation of crystal form-controlled and highly crystalline $BiVO_4$ powder from layered vanadates at room temperature and its photocatalytic and photophysical properties. *Journal of the American Chemical Society, 121*, 11459–11467.

18. Maeda, K., et al. (2005). GaN:ZnO solid solution as a photocatalyst for visible-light-driven overall water splitting. *Journal of the American Chemical Society, 127*, 8286–8287.

19. Bard, A. J., Memming, R., & Miller, B. (1991). Terminology in semiconductor electrochemistry and photoelectochemical conversion—Recommendations. *Pure and Applied Chemistry, 63*, 569–596.

20. Zhang, W., Tang, J., & Ye, J. (2007). Structural, photocatalytic, and photophysical properties of perovskite $MSnO_3$ (M = Ca, Sr, and Ba) photocatalysts. *Journal of Materials Research, 22*, 1859–1871.

21. Bisquert, J. (2008). Physical electrochemistry of nanostructured devices. *Physical Chemistry Chemical Physics, 10*, 49–72.

22. Grey, I. E. (2004) Electrochemical Water Splitting—A review of published research. *CSIRO Minerals: 2004*.

23. Kudo, A., & Miseki, Y. (2009). Heterogeneous photocatalyst materials for water splitting. *Chemical Society Reviews, 38*, 253–278.

24. Park, H. G., & Holt, J. K. (2010). Recent advances in nanoelectrode architecture for photochemical hydrogen production. *Energy and Environmental Science, 3*, 1028–1036.

25. Rajeshwar, K. (2007). Hydrogen generation at irradiated oxide semiconductor—solution interfaces. *Journal of Applied Electrochemistry, 37*, 765–787.

26. Jing, D., et al. (2010). Efficient solar hydrogen production by photocatalytic water splitting: From fundamental study to pilot demonstration. *International Journal of Hydrogen Energy, 35*, 7087–7097.

27. Varghese, O. K., & Grimes, C. A. (2008). Appropriate strategies for determining the photoconversion efficiency of water photoelectrolysis cells: A review with examples using titania nanotube array photoanodes. *Solar Energy Materials and Solar Cells, 92*, 374–384.

28. Archer, M. D. (2008). *Nanostructured and photoelectrochemical systems for solar photon conversion : Series on photoconversion of solar energy*.

29. Vayssieres, L. (2009). *On solar hydrogen and nanotechnology*. Chichester: Wiley.

30. Pendlebury, S. R., et al. (2011). Dynamics of photogenerated holes in nanocrystalline á-Fe_2O_3 electrodes for water oxidation probed by transient absorption spectroscopy. *Chemical Communications, 47*.

31. Yu, J., & Kudo, A. (2006). Effects of structural variation on the photocatalytic performance of hydrothermally synthesized $BiVO_4$. *Advanced Functional Materials, 16*, 2163–2169.

32. Parida, K. M., Reddy, K. H., Martha, S., Das, D. P., & Biswal, N. (2010). Fabrication of nanocrystalline $LaFeO_3$: An efficient sol-gel auto-combustion assisted visible light responsive photocatalyst for water decomposition. *International Journal of Hydrogen Energy, 35*, 12161–12168.

33. Allen, L. C. (1989). Electronegativity is the average one-electron energy of the valence-shell electrons in ground-state free atoms. *Journal of the American Chemical Society, 111*, 9003–9014.

34. Mills, A., & Valenzuela, M. A. (2004). Photo-oxidation of water sensitized by TiO_2 and WO_3 in presence of different electron acceptors. *Revista Mexicana de Fisica, 50*, 287–296.

35. Mills, A., & Valenzuela, M. A. (2004). The photo-oxidation of water by sodium persulfate, and other electron acceptors, sensitised by TiO_2. *Journal of Photochemistry and Photobiology A: Chemistry, 165*, 25–34.

36. Bamwenda, G. R., Tsubota, S., Nakamura, T., & Haruta, M. (1995). Photoassisted hydrogen production from a water-ethanol solution: A comparison of activities of Au-TiO_2 and Pt-TiO_2. *Journal of Photochemistry and Photobiology A: Chemistry, 89*, 177–189.

37. Li Puma, G. (2003). Modeling of thin-film slurry photocatalytic reactors affected by radiation scattering. *Environmental Science and Technology, 37*, 5783–5791.

38. Serpone, N., Terzian, R., Lawless, D., Kennepohl, P., & Sauvé, G. (1993). On the usage of turnover numbers and quantum yields in heterogeneous photocatalysis. *Journal of Photochemistry and Photobiology A: Chemistry, 73*, 11–16.

39. Weber, M. F., & Dignam, M. J. (1984). Efficiency of splitting water with semiconducting photoelectrodes. *Journal of the Electrochemical Society, 131,* 1258–1265.
40. Bolton, J. R., Haught, A. F. & Ross, R. T. (1981). Photochemical energy storage: An analysis of limits. *Photochemical Conversion and Storage of Solar Energy,* 297–330.
41. Bolton, J. R., Strickler, S. J., & Connolly, J. S. (1985). Limiting and realizable efficiencies of solar photolysis of water. *Nature, 316,* 495–500.
42. Tang, J., Cowan, A. J., Durrant, J. R., & Klug, D. R. (2011). Mechanism of O_2 production from water splitting: Nature of charge carriers in nitrogen doped nanocrystalline TiO_2 films and factors limiting O_2 production. *The Journal of Physical Chemistry C, 115,* 3143–3150.
43. Tauc, J. (1968). Optical properties and electronic structure of amorphous Ge and Si. *Materials Research Bulletin, 3,* 37–46.
44. Tsunekawa, S., Sahara, R., Kawazoe, Y. & Kasuya, A. (2000). Origin of the blue shift in ultraviolet absorption spectra of nanocrystalline $CeO_{[2-x]}$ particles. *Materials Transactions, 233.*
45. Kadoya, K., Matsunaga, N., & Nagashima, A. (1985). Viscosity and thermal conductivity of dry air in the gaseous phase. *Journal of Physical and Chemical Reference Data, 14,* 947–970.
46. Bragg, W. L. (1934). *The crystalline state* (Vol. I). New York: The Macmillan Company.
47. Hart, M. (1975). X-ray diffraction by L. V. Azaroff, R. Kaplow, N. Kato, R. J. Weiss, A. J. C. Wilson and R. A. Young. *Acta Crystallographica Section A, 31,* 878.
48. Bueno-Ferrer, C., Parres-Esclapez, S., Lozano-Castelló, D., & Bueno-López, A. (2010). Relationship between surface area and crystal size of pure and doped cerium oxides. *Journal of Rare Earths, 28,* 647–653.
49. Egerton, R. F. (2005). *Physical principles of electron microscopy: An introduction to TEM, SEM, and AEM.* New York: Springer Science + Business Media.
50. Brunauer, S., Emmett, P. H., & Teller, E. (1938). Adsorption of gases in multimolecular layers. *Journal of the American Chemical Society, 60,* 309–319.
51. Hanaor, D., Michelazzi, M., Leonelli, C., & Sorrell, C. C. (2012). The effects of carboxylic acids on the aqueous dispersion and electrophoretic deposition of ZrO_2. *Journal of the European Ceramic Society, 32,* 235–244.
52. Mirabella, F. M. (1992). *Internal reflection spectroscopy: Theory and applications* (Vol. 15). West Palm Beach: CRC Press.
53. White, R. (1989). *Chromatography/Fourier Transform Infrared Spectroscopy and its Applications.* London: Taylor & Francis.
54. Greener, J., Abbasi, B., & Kumacheva, E. (2010). Attenuated total reflection Fourier transform infrared spectroscopy for on-chip monitoring of solute concentrations. *Lab on a Chip, 10,* 1561–1566.
55. Smith, E. & Dent, G. (2005). *Modern Raman spectroscopy: A practical approach.* Chichester: Wiley.
56. Bernath, P. F. *Spectra of atoms and molecules.* (Oxford University Press, 2005).
57. Hollander, J. M., & Jolly, W. L. (1970). X-ray photoelectron spectroscopy. *Accounts of Chemical Research, 3,* 193–200.
58. Wagner, C. & Muilenberg, G. (1979). *Handbook of X-ray photoelectron spectroscopy.* Boca Raton, FL: Perkin-Elmer.
59. Moulder, J. F., Stickle, W. F., Sobol, P. E., & Bomben, K. D. (1992). *Handbook of X-ray photoelectron spectroscopy: A reference book of standard spectra for identification and interpretation of XPS Data.* Boca Raton, FL: Perkin-Elmer.
60. Asahi, R., Morikawa, T., Ohwaki, T., Aoki, K., & Taga, Y. (2001). Visible-light photocatalysis in nitrogen-doped titanium oxides. *Science, 293,* 269–271.
61. Wang, J., et al. (2009). Origin of photocatalytic activity of nitrogen-doped TiO_2 nanobelts. *Journal of the American Chemical Society, 131,* 12290–12297.
62. Ishikawa, Y., Matsumoto, Y., Nishida, Y., Taniguchi, S., & Watanabe, J. (2003). Surface treatment of silicon carbide using $TiO_2(IV)$ photocatalyst. *Journal of the American Chemical Society, 125,* 6558–6562.

63. Kawasaki, S., et al. (2012). Epitaxial Rh-doped $SrTiO_3$ thin film photocathode for water splitting under visible light irradiation. *Applied Physics Letters, 101.* doi:http://dx.doi.org/10.1063/1061.4738371.

64. Holmes, F. L. (1963). Elementary analysis and the origins of physiological chemistry. *Isis, 50*–81.

65. Pavia, D., Lampman, G., Kriz, G. & Vyvyan, J. (2008). *Introduction to spectroscopy.* Stamford: Cengage Learning.

66. Li, K., Martin, D. J., & Tang, J. (2011). Conversion of solar energy to fuels by inorganic heterogeneous systems. *Chinese Journal of Catalysis, 32,* 879–890.

67. Pei, D., & Luan, J. (2012). Development of visible light-responsive sensitized photocatalysts. *International Journal of Photoenergy, 2012,* 13.

68. Khaselev, O., & Turner, J. A. (1998). A monolithic photovoltaic-photoelectrochemical device for hydrogen production via water splitting. *Science, 280,* 425–427.

69. Yamaguti, K., & Sato, S. (1985). Pressure dependence of the rate and stoichiometry of water photolysis over platinized TiO_2 (anatase and rutile) catalysts. *The Journal of Physical Chemistry, 89,* 5510–5513.

70. Selli, E., et al. (2007). A photocatalytic water splitting device for separate hydrogen and oxygen evolution. *Chemical Communications,* 5022–5024.

71. Duonghong, D. (1981). Dynamics of light-induced water cleavage in colloidal systems. *Journal of the American Chemical Society, 103,* 4685.

72. Sato, S., & White, J. M. (1981). Photoassisted hydrogen production from titania and water. *The Journal of Physical Chemistry, 85,* 592–594.

73. Fu, N., Wu, Y., Jin, Z., & Lu, G. (2010). Structural-Dependent photoactivities of TiO_2 nanoribbon for visible-light-induced H_2 evolution: The roles of nanocavities and alternate structures. *Langmuir, 26,* 447–455.

74. Li, Y., Xie, Y., Peng, S., Lu, G., & Li, S. (2006). Photocatalytic hydrogen generation in the presence of chloroacetic acids over Pt/TiO_2. *Chemosphere, 63,* 1312–1318.

75. Domen, K. (1986). Photocatalytic decomposition of water into hydrogen and oxygen over nickel (II) oxide-strontium titanate ($SrTiO_3$) powder. *Journal of Physical Chemistry, 90,* 292.

76. Domen, K., Kudo, A., & Onishi, T. (1986). Mechanism of photocatalytic decomposition of water into H_2 and O_2 over NiO-$SrTiO_3$. *Journal of Catalysis, 102,* 92–98.

77. Domen, K., Naito, S., Onishi, T., & Tamaru, K. (1982). Photocatalytic decomposition of liquid water on a NiO-$SrTiO_3$ catalyst. *Chemical Physics Letters, 92,* 433–434.

78. Kudo, A., et al. (1989). Nickel-loaded $K_4Nb_6O_{17}$ photocatalyst in the decomposition of H_2O into H_2 and O_2: Structure and reaction mechanism. *Journal of Catalysis, 120,* 337–352.

79. Kudo, A., et al. (1988). Photocatalytic decomposition of water over NiO-$K_4Nb_6O_{17}$ catalyst. *Journal of Catalysis, 111,* 67–76.

80. Ji, S. M., & Min, J. (2005). Photocatalytic hydrogen production from water methanol mixtures using N-doped $Sr_2Nb_2O_7$ under visible light irradiation: Effects of catalyst structure. *Physical chemistry chemical physics, 7,* 1315.

81. Chang, S., & Doong, R. (2004). The effect of chemical states of dopants on the microstructures and band gaps of metal-doped ZrO_2 thin films at different temperatures. *The Journal of Physical Chemistry B, 108,* 18098–18103.

82. Sayama, K., & Arakawa, H. (1993). Photocatalytic decomposition of water and photocatalytic reduction of carbon dioxide over zirconia catalyst. *The Journal of Physical Chemistry, 97,* 531–533.

83. Jiang, L., Wang, Q., Li, C., Yuan, J., & Shangguan, W. (2010). ZrW_2O_8 photocatalyst and its visible-light sensitization via sulfur anion doping for water splitting. *International Journal of Hydrogen Energy, 35,* 7043–7050.

84. Sayama, K. (1996). Photocatalytic water splitting on nickel intercalated $A_4Ta_xNb_{6-x}O_{17}$ (A = K, Rb). *Catalysis Today, 28,* 175.

85. Kato, H. (1998). New tantalate photocatalysts for water decomposition into H_2 and O_2. *Chemical Physics Letters, 295*, 487.
86. Kato, H. (2001). Water splitting into H_2 and O_2 on alkali tantalate photocatalysts $ATaO_3$ (A = Li, Na, and K). *The Journal of Physical Chemistry B, 105*, 4285.
87. Kato, H., Asakura, K., & Kudo, A. (2003). Highly efficient water splitting into H_2 and O_2 over lanthanum-doped $NaTaO_3$ photocatalysts with high crystallinity and surface nanostructure. *Journal of the American Chemical Society, 125*, 3082–3089.
88. Mitsui, C., Nishiguchi, H., Fukamachi, K., Ishihara, T., & Takita, Y. (1999). Photocatalytic decomposition of pure water over NiO supported on $KTa(M)O_3$ (M = Ti^{4+}, Hf^{4+}, Zr^{4+}) perovskite oxide. *Chemistry Letters, 28*, 1327–1328.
89. Yoshino, M. (2002). Polymerizable complex synthesis of pure $Sr_2Nb_xTa_{2-x}O_7$ solid solutions with high photocatalytic activities for water decomposition into H_2 and O_2. *Chemistry of Materials, 14*, 3369.
90. Domen, K., et al. (1986). Photodecomposition of water and hydrogen evolution from aqueous methanol solution over novel niobate photocatalysts. *Journal of the Chemical Society, Chemical Communications*, 356–357.
91. Wiegelm, M., Emond, M. H. J., Stobbe, E. R. & Blasse, G. (1994). Luminescence of alkali tantalates and niobates. *Journal of Physics and Chemistry of Solids, 55*, 773–778.
92. Sato, J., Kobayashi, H., & Inoue, Y. (2003). Photocatalytic activity for water decomposition of indates with octahedrally coordinated d10 configuration. II. Roles of geometric and electronic structures. *The Journal of Physical Chemistry B, 107*, 7970–7975.
93. Sato, J. (2003). Photocatalytic activities for water decomposition of RuO_2-loaded $AInO_2$ (A = Li, Na) with d10 configuration. *Journal of Photochemistry and Photobiology. A, Chemistry, 158*, 139.
94. Zhang, W. F., Tang, J., & Ye, J. (2006). Photoluminescence and photocatalytic properties of $SrSnO_3$ perovskite. *Chemical Physics Letters, 418*, 174–178.
95. Chen, D., & Ye, J. (2007). SrSnO3 nanostructures: Synthesis, characterization, and photocatalytic properties. *Chemistry of Materials, 19*, 4585–4591.
96. Sato, J., Saito, N., Nishiyama, H., & Inoue, Y. (2002). Photocatalytic water decomposition by RuO_2-loaded antimonates, $M_2Sb_2O_7$ (M = Ca, Sr), $CaSb_2O_6$ and $NaSbO_3$, with d10 configuration. *Journal of Photochemistry and Photobiology A: Chemistry, 148*, 85–89.
97. Yanagida, T. (2004). Photocatalytic decomposition of H_2O into H_2 and O_2 over Ga_2O_3 loaded with NiO. *Chemistry Letters, 33*, 726.
98. Sakata, Y. (2008). Effect of metal ion addition in a Ni supported Ga_2O_3 photocatalyst on the photocatalytic overall splitting of H_2O. *Catalysis Letters, 125*, 22.
99. Ikarashi, K. (2002). Photocatalysis for Water Decomposition by RuO_2-Dispersed $ZnGa_2O_4$ with d10 Configuration. *The Journal of Physical Chemistry B, 106*, 9048.
100. Yanagida, S. (1982). Photocatalytic hydrogen evolution from water using zinc sulfide and sacrificial electron donors. *Chemistry Letters, 11*, 1069.
101. Reber, J. F., & Meier, K. (1984). Photochemical production of hydrogen with zinc sulfide suspensions. *The Journal of Physical Chemistry, 88*, 5903–5913.
102. Kobayakawa, K. (1992). Photocatalytic activity of $CuInS_2$ and $CuIn_5S_8$. *Electrochimica Acta, 37*, 465.
103. Sato, J., et al. (2005). RuO_2-loaded beta-Ge_3N_4 as a non-oxide photocatalyst for overall water splitting. *Journal of the American Chemical Society, 127*, 4150–4151.
104. Lee, Y. G., et al. (2006). Effect of high-pressure ammonia treatment on the activity of Ge_3N_4 photocatalyst for overall water splitting. *Journal of Physical Chemistry B, 110*, 17563–17569.
105. Maeda, K., Saito, N., Inoue, Y., & Domen, K. (2007). Dependence of activity and stability of germanium nitride powder for photocatalytic overall water splitting on structural properties. *Chemistry of Materials, 19*, 4092–4097.

106. Maeda, K., Teramura, K., Saito, N., Inoue, Y., & Domen, K. (2007). Photocatalytic overall water splitting on gallium nitride powder. *Bulletin of the Chemical Society of Japan, 80,* 1004–1010.
107. Arai, N., et al. (2006). Overall water splitting by RuO$_2$-dispersed divalent-ion-doped GaN photocatalysts with d(10) electronic configuration. *Chemistry Letters, 35,* 796–797.
108. Wei, S. H., & Zunger, A. (1988). Role of metal d states in II-VI semiconductors. *Physical Review B, 37,* 8958.
109. Kawai, T. & Sakata, T. (1980). Photocatalytic hydrogen production from liquid methanol and water. *Journal of the Chemical Society, Chemical Communications,* 694–695.
110. Kawai, T. & Sakata, T. (1980). Conversion of carbohydrate into hydrogen fuel by a photo-catalytic process. *Nature, 286*(5772), 474–476.
111. Sakata, T., & Kawai, T. (1981). Heterogeneous photocatalytic production of hydrogen and methane from ethanol and water. *Chemical Physics Letters, 80,* 341–344.
112. Borgarello, E., Kiwi, J., Graetzel, M., Pelizzetti, E., & Visca, M. (1982). Visible light induced water cleavage in colloidal solutions of chromium-doped titanium dioxide particles. *Journal of the American Chemical Society, 104,* 2996–3002.
113. Chen, X., & Mao, S. S. (2007). Titanium dioxide nanomaterials: Synthesis, properties, mod-ifications, and applications. *Chemical Reviews, 107,* 2891–2959.
114. Ji, P., et al. (2010). Recent advances in visible light-responsive titanium oxide-based photo-catalysts. *Research on Chemical Intermediates, 36,* 327–347.
115. Leung, D. Y. C., et al. (2010). Hydrogen production over titania-based photocatalysts. *ChemSusChem, 3,* 681–694.
116. Anpo, M., et al. (2001). The design and development of second-generation titanium oxide photocatalysts able to operate under visible light irradiation by applying a metal ion-implan-tation method. *Research on Chemical Intermediates, 27,* 459–467.
117. Choi, W., Termin, A., & Hoffmann, M. R. (1994). The role of metal ion dopants in quan-tum-sized TiO$_2$: Correlation between photoreactivity and charge carrier recombination dynamics. *The Journal of Physical Chemistry, 98,* 13669–13679.
118. Rengaraj, S., & Li, X. (2006). Enhanced photocatalytic activity of TiO$_2$ by doping with Ag for degradation of 2,4,6-trichlorophenol in aqueous suspension. *Journal of Molecular Catalysis A: Chemical, 243,* 60–67.
119. Kim, S., Hwang, S.-J., & Choi, W. (2005). Visible light active platinum-ion-doped TiO$_2$ photocatalyst. *The Journal of Physical Chemistry B, 109,* 24260–24267.
120. Umebayashi, T., Yamaki, T., Itoh, H., & Asai, K. (2002). Band gap narrowing of titanium dioxide by sulfur doping. *Applied Physics Letters, 81,* 454–456.
121. Choi, Y., Umebayashi, T., & Yoshikawa, M. (2004). Fabrication and characterization of C-doped anatase TiO$_2$ photocatalysts. *Journal of Materials Science, 39,* 1837–1839.
122. Reyes-Garcia, E. A., Sun, Y., & Raftery, D. (2007). Solid-state characterization of the nuclear and electronic environments in a boron-fluoride co-doped TiO$_2$ visible-light photo-catalyst. *The Journal of Physical Chemistry C, 111,* 17146–17154.
123. Konta, R., Ishii, T., Kato, H., & Kudo, A. (2004). Photocatalytic activities of noble metal ion doped SrTiO$_3$ under visible light irradiation. *The Journal of Physical Chemistry B, 108,* 8992–8995.
124. Nishimoto, S., Matsuda, M., & Miyake, M. (2006). Photocatalytic activities of Rh-doped CaTiO$_3$ under visible light irradiation. *Chemistry Letters, 35,* 308–309.
125. Okazaki, Y., Mishima, T., Nishimoto, S., Matsuda, M., & Miyake, M. (2008). Photocatalytic activity of Ca$_3$Ti$_2$O$_7$ layered-perovskite doped with Rh under visible light irradiation. *Materials Letters, 62,* 3337–3340.
126. Zou, Z., Ye, J., Sayama, K., & Arakawa, H. (2001). Direct splitting of water under visible light irradiation with an oxide semiconductor photocatalyst. *Nature, 414,* 625–627.
127. Iwase, A., Saito, K., & Kudo, A. (2009). Sensitization of NaMO$_3$ (M: Nb and Ta) photocata-lysts with wide band gaps to visible light by Ir doping. *Bulletin of the Chemical Society of Japan, 82,* 514–518.

128. Hara, M. (2003). TaON and Ta$_3$N$_5$ as new visible light driven photocatalysts. *Catalysis Today, 78*, 555.

129. Hara, M., Nunoshige, J., Takata, T., Kondo, J. N. & Domen, K. (2003). Unusual enhancement of H$_2$ evolution by Ru on TaON photocatalyst under visible light irradiation. *Chemical Communications*, 3000–3001.

130. Liu, M., et al. (2004). Water reduction and oxidation on Pt-Ru/Y$_2$Ta$_2$O$_5$N$_2$ catalyst under visible light irradiation. *Chemical Communications*, 2192–2193.

131. Yamasita, D., Takata, T., Hara, M., Kondo, J. N., & Domen, K. (2004). Recent progress of visible-light-driven heterogeneous photocatalysts for overall water splitting. *Solid State Ionics, 172*, 591–595.

132. Maeda, K., Teramura, K., & Domen, K. (2008). Effect of post-calcination on photocatalytic activity of (Ga$_{1-x}$Zn$_x$)(N$_{1-x}$O$_x$) solid solution for overall water splitting under visible light. *Journal of Catalysis, 254*, 198–204.

133. Maeda, K., et al. (2006). Photocatalyst releasing hydrogen from water. *Nature, 440*, 295–295.

134. Tang, J. (2005). Correlation of crystal structures and electronic structures and photocatalytic properties of the W-containing oxides. *Journal of Materials Chemistry, 15*, 4246.

135. Tang, J. (2003). Photophysical and photocatalytic properties of AgInW$_2$O$_8$. *the Journal of physical Chemistry B, 107*, 14265.

136. Kudo, A. (1998). Photocatalytic O$_2$ evolution under visible light irradiation on BiVO$_4$ in aqueous AgNO$_3$ solution. *Catalysis Letters, 53*, 229.

137. Darwent, J. R. (1981). H$_2$ production photosensitized by aqueous semiconductor dispersions. *Journal of the Chemical Society, Faraday Transactions 2: Molecular and Chemical Physics, 77*, 1703–1709.

138. Darwent, J. R. & Porter, G. (1981). Photochemical hydrogen production using cadmium sulphide suspensions in aerated water. *Journal of the Chemical Society, Chemical Communications*, 145–146.

139. Yan, H., et al. (2009). Visible-light-driven hydrogen production with extremely high quantum efficiency on Pt–PdS/CdS photocatalyst. *Journal of Catalysis, 266*, 165–168.

140. Ma, G., et al. (2008). Direct splitting of H$_2$S into H$_2$ and S on CdS-based photocatalyst under visible light irradiation. *Journal of Catalysis, 260*, 134–140.

141. Tabata, M., et al. (2010). Photocatalytic hydrogen evolution from water using copper gallium sulfide under visible-light irradiation. *The Journal of Physical Chemistry C, 114*, 11215–11220.

142. Zong, X., et al. (2008). Enhancement of photocatalytic H$_2$ evolution on CdS by loading MoS$_2$ as cocatalyst under visible light irradiation. *Journal of the American Chemical Society, 130*, 7176–7177.

143. Kalyanasundaram, K., Borgarello, E., Duonghong, D., & Grätzel, M. (1981). Cleavage of water by visible-light irradiation of colloidal CdS solutions; Inhibition of photocorrosion by RuO$_2$. *Angewandte Chemie, International Edition in English, 20*, 987–988.

144. Youn, H. C., Baral, S., & Fendler, J. H. (1988). Dihexadecyl phosphate, vesicle-stabilized and insitu generated mixed CdS and ZnS semiconductor particles—Preparation and utilization for photosensitized charge separation and hydrogen generation. *Journal of Physical Chemistry, 92*, 6320–6327.

145. Sabate, J., Cerveramarch, S., Simarro, R., & Gimenez, J. (1990). A comparative-study of semiconductor photocatalysts for hydrogen-production by visible-light using different sacrificial substrates in aqueous-media. *International Journal of Hydrogen Energy, 15*, 115–124.

146. Xing, C. J., Zhang, Y. J., Yan, W., & Guo, L. J. (2006). And structure-controlled solid solution of Cd$_{1-x}$Zn$_x$S photocatalyst for hydrogen production by water splitting. *International Journal of Hydrogen Energy, 31*, 2018–2024.

147. Rufus, I. B., Viswanathan, B., Ramakrishnan, V., & Kuriacose, J. C. (1995). Cadmium-sulfide with iridium sulfide and platinum sulfide deposits as a photocatalyst for the decomposition of aqueous sulfide. *Journal of Photochemistry and Photobiology A, 91*, 63–66.

148. Umezawa, N., Shuxin, O., & Ye, J. (2011). Theoretical study of high photocatalytic performance of Ag_3PO_4. *Physical Review B, 83*, 035202.
149. Bi, Y., Ouyang, S., Umezawa, N., Cao, J., & Ye, J. (2011). Facet effect of single-crystalline Ag_3PO_4 sub-microcrystals on photocatalytic properties. *Journal of the American Chemical Society, 133*, 6490–6492.
150. Bi, Y., Ouyang, S., Cao, J., & Ye, J. (2011). Facile synthesis of rhombic dodecahedral AgX/Ag_3PO_4 (X = Cl, Br, I) heterocrystals with enhanced photocatalytic properties and stabilities. *Physical Chemistry Chemical Physics, 13*, 10071–10075.
151. Wang, X., et al. (2008). A metal-free polymeric photocatalyst for hydrogen production from water under visible light. *Nature Materials, 8*, 76–80.
152. Liebig, J. (1834). Ueber Einige Stickstoff-Verbindungen. *Annalen der Pharmacie, 10*, 1–47.
153. Franklin, E. C. (1922). The ammono carbonic acids. *Journal of the American Chemical Society, 44*, 486–509.
154. Pauling, L., & Sturdivant, J. H. (1937). The structure of cyameluric acid, hydromelonic acid and related substances. *Proceedings of the National Academy of Sciences, 23*, 615–620.
155. Lotsch, B. V., et al. (2007). Unmasking Melon by a complementary approach employing electron diffraction, solid-state NMR spectroscopy, and theoretical calculations—Structural characterization of a carbon nitride polymer. *Chemistry – A European Journal, 13*, 4969–4980.
156. Kroke, E., & Schwarz, M. (2004). Novel group 14 nitrides. *Coordination Chemistry Reviews, 248*, 493–532.
157. Kroke, E., et al. (2002). Tri-s-triazine derivatives. Part I. From trichloro-tri-s-triazine to graphitic C_3N_4 structures. *New Journal of Chemistry, 26*, 508–512.
158. Dong, F., Sun, Y., Wu, L., Fu, M., & Wu, Z. (2012). Facile transformation of low cost thiourea into nitrogen-rich graphitic carbon nitride nanocatalyst with high visible light photocatalytic performance. *Catalysis Science & Technology, 2*, 1332–1335.
159. Dong, F., et al. (2011). Efficient synthesis of polymeric g-C_3N_4 layered materials as novel efficient visible light driven photocatalysts. *Journal of Materials Chemistry, 21*, 15171–15174.
160. Yue, B., Li, Q., Iwai, H., Kako, T., & Ye, J. (2011). Hydrogen production using zinc-doped carbon nitride catalyst irradiated with visible light. *Science and Technology of Advanced Materials, 12*, 034401.
161. Jorge, A. B., et al. (2013). H_2 and O_2 evolution from water half-splitting reactions by graphitic carbon nitride materials. *The Journal of Physical Chemistry C, 117*, 7178–7185.
162. Ge, L., Han, C., Xiao, X., Guo, L. & Li, Y. (2013) Enhanced visible light photocatalytic hydrogen evolution of sulfur-doped polymeric g-C_3N_4 photocatalysts. *Materials Research Bulletin, 48*, 3919–3925.
163. Schwinghammer, K., et al. (2013). Triazine-based carbon nitrides for visible-light-driven hydrogen evolution. *Angewandte Chemie International Edition, 52*, 2435–2439.
164. Xiang, Q., Yu, J., & Jaroniec, M. (2011). Preparation and enhanced visible-light photocatalytic H_2-production activity of graphene/C_3N_4 composites. *The Journal of Physical Chemistry C, 115*, 7355–7363.
165. Wang, X., et al. (2009). Polymer semiconductors for artificial photosynthesis: Hydrogen evolution by mesoporous graphitic carbon nitride with visible light. *Journal of the American Chemical Society, 131*, 1680–1681.
166. Xu, J., Li, Y., Peng, S., Lu, G., & Li, S. (2013). Eosin Y-sensitized graphitic carbon nitride fabricated by heating urea for visible light photocatalytic hydrogen evolution: The effect of the pyrolysis temperature of urea. *Physical Chemistry Chemical Physics, 15*, 7657–7665.
167. Wang, Y., Zhang, J., Wang, X., Antonietti, M., & Li, H. (2010). Boron-and fluorine-containing mesoporous carbon nitride polymers: Metal-free catalysts for cyclohexane oxidation. *Angewandte Chemie International Edition, 49*, 3356–3359.
168. Liu, G., et al. (2010). Unique electronic structure induced high photoreactivity of sulfur-doped graphitic C_3N_4. *Journal of the American Chemical Society, 132*, 11642–11648.

169. Bard, A. J. (1979). Photoelectrochemistry and heterogeneous photo-catalysis at semiconductors. *Journal of Photochemistry, 10*, 59–75.
170. Kato, H., Hori, M., Konta, R., Shimodaira, Y., & Kudo, A. (2004). Construction of Z-scheme type heterogeneous photocatalysis systems for water splitting into H_2 and O_2 under visible light irradiation. *Chemistry Letters, 33*, 1348–1349.
171. Bowker, M. (2011). Sustainable hydrogen production by the application of ambient temperature photocatalysis. *Green Chemistry, 13*, 2235–2246.
172. Martin, D. J., Reardon, P. J. T., Handoko, A. D. & Tang, J. (2014). Visible light-driven pure water splitting by a nature-inspired organic semiconductor system. *Journal of the American Chemical Society (In review)*.
173. Darwent, J. R., & Mills, A. (1982). Photo-oxidation of water sensitized by WO_3 powder. *Journal of the Chemical Society, Faraday Transactions 2: Molecular and Chemical Physics, 78*, 359–367.
174. Abe, R., Sayama, K., Domen, K., & Arakawa, H. (2001). A new type of water splitting system composed of two different TiO_2 photocatalysts (anatase, rutile) and a IO_3^-/I^- shuttle redox mediator. *Chemical Physics Letters, 344*, 339–344.
175. Kim, Y. I., Salim, S., Huq, M. J., & Mallouk, T. E. (1991). Visible-light photolysis of hydrogen iodide using sensitized layered semiconductor particles. *Journal of the American Chemical Society, 113*, 9561–9563.
176. Sayama, K., Mukasa, K., Abe, R., Abe, Y. & Arakawa, H. Stoichiometric water splitting into H_2 and O_2 using a mixture of two different photocatalysts and an IO_3^-/I^- shuttle redox mediator under visible light irradiation. *Chemical Communications*, 2416–2417, (2001).
177. Ohno, T., Haga, D., Fujihara, K., Kaizaki, K., & Matsumura, M. (1997). Unique effects of iron (III) ions on photocatalytic and photoelectrochemical properties of titanium dioxide. *The Journal of Physical Chemistry B, 101*, 6415–6419.
178. Abe, R., Sayama, K., & Sugihara, H. (2005). Development of new photocatalytic water Splitting into H_2 and O_2 using two different semiconductor photocatalysts and a shuttle redox mediator IO_3^-/I^-. *The Journal of Physical Chemistry B, 109*, 16052–16061.
179. Tabata, M., et al. (2010). Modified Ta_3N_5 powder as a photocatalyst for O_2 evolution in a two-step water splitting system with an iodate/iodide shuttle redox mediator under visible light. *Langmuir, 26*, 9161–9165.
180. Maeda, K., Higashi, M., Lu, D., Abe, R., & Domen, K. (2010). Efficient nonsacrificial water splitting through two-step photoexcitation by visible light using a modified oxynitride as a hydrogen evolution photocatalyst. *Journal of the American Chemical Society, 132*, 5858–5868.
181. Maeda, K. (2013). Z-scheme water splitting using two different semiconductor photocatalysts. *ACS Catalysis, 3*, 1486–1503.
182. Sayama, K., et al. (1997). Photocatalytic decomposition of water into H_2 and O_2 by a two-step photoexcitation reaction using a WO_3 suspension catalyst and an Fe^{3+}/Fe^{2+} redox system. *Chemical Physics Letters, 277*, 387–391.
183. Hara, S., & Irie, H. (2012). Band structure controls of $SrTiO_3$ towards two-step overall water splitting. *Applied Catalysis, B: Environmental, 115*, 330–335.
184. Hara, S., et al. (2012). Hydrogen and oxygen evolution photocatalysts synthesized from strontium titanate by controlled doping and their performance in two-step overall water splitting under visible light. *The Journal of Physical Chemistry C, 116*, 17458–17463.
185. Sayama, K., Abe, R., Arakawa, H., & Sugihara, H. (2006). Decomposition of water into H_2 and O_2 by a two-step photoexcitation reaction over a Pt–TiO_2 photocatalyst in $NaNO_2$ and Na_2CO_3 aqueous solution. *Catalysis Communications, 7*, 96–99.
186. Sasaki, Y., Iwase, A., Kato, H., & Kudo, A. (2008). The effect of co-catalyst for Z-scheme photocatalysis systems with an Fe^{3+}/Fe^{2+} electron mediator on overall water splitting under visible light irradiation. *Journal of Catalysis, 259*, 133–137.
187. Sasaki, Y., Nemoto, H., Saito, K., & Kudo, A. (2009). Solar water splitting using powdered photocatalysts driven by Z-schematic interparticle electron transfer without an electron mediator. *The Journal of Physical Chemistry C, 113*, 17536–17542.

188. Iwase, A., Ng, Y. H., Ishiguro, Y., Kudo, A., & Amal, R. (2011). Reduced graphene oxide as a solid-state electron mediator in Z-scheme photocatalytic water splitting under visible light. *Journal of the American Chemical Society, 133*, 11054–11057.

189. Abe, R., Takata, T., Sugihara, H. & Domen, K. Photocatalytic overall water splitting under visible light by TaON and WO_3 with an IO_3^-/I^- shuttle redox mediator. *Chemical Communications*, 3829–3831, (2005).

190. Abe, R., Higashi, M., & Domen, K. (2011). Overall water splitting under visible light through a two-step photoexcitation between TaON and WO_3 in the presence of an iodate-iodide shuttle redox mediator. *ChemSusChem, 4*, 228–237.

191. Higashi, M., Abe, R., Ishikawa, A., Takata, T., & Ohtani, B. (2008). Z-scheme overall water splitting on modified-TaON photocatalysts under visible light (500 nm). *Chemistry Letters, 37*, 138–139.

192. Maeda, K., et al. (2008). Surface modification of TaON with monoclinic ZrO_2 to produce a composite photocatalyst with enhanced hydrogen evolution activity under visible light. *Bulletin of the Chemical Society of Japan, 81*, 927–937.

193. Maeda, K., Abe, R., & Domen, K. (2011). Role and function of ruthenium species as promoters with TaON-based photocatalysts for oxygen evolution in two-step water splitting under visible light. *The Journal of Physical Chemistry C, 115*, 3057–3064.

194. Hodes, G., Albu-Yaron, A., Decker, F., & Motisuke, P. (1987). Three-dimensional quantum-size effect in chemically deposited cadmium selenide films. *Physical Review B, 36*, 4215.

195. Furukawa, S., & Miyasato, T. (1988). Quantum size effects on the optical band gap of microcrystalline Si:H. *Physical Review B, 38*, 5726.

196. Saremi-Yarahmadi, S., Vaidhyanathan, B., & Wijayantha, K. G. U. (2010). Microwave-assisted low temperature fabrication of nanostructured alpha-Fe_2O_3 electrodes for solar-driven hydrogen generation. *International Journal of Hydrogen Energy, 35*, 10155–10165.

197. Ziegler, J., Merkulov, A., Grabolle, M., Resch-Genger, U., & Nann, T. (2007). High-quality ZnS shells for CdSe nanoparticles: Rapid microwave synthesis. *Langmuir, 23*, 7751–7759.

198. Wang, X., Zhuang, J., Peng, Q., & Li, Y. D. (2005). A general strategy for nanocrystal synthesis. *Nature, 437*, 121–124.

199. Song, H., Cai, P., Huabing, Y., & Yan, C. (2007). Hydrothermal synthesis of flaky crystallized $La_2Ti_2O_7$ for producing hydrogen from photocatalytic water splitting. *Catalysis Letters, 113*, 54–58.

200. Haynes, C. L., McFarland, A. D. & Duyne, R. P. V. (2005). Surface-enhanced raman spectroscopy. *Analytical Chemistry, 77*, 338 A–346 A.

201. Wiley, B. J., et al. (2006). Synthesis and electrical characterization of silver nanobeams. *Nano Letters, 6*, 2273–2278.

202. Aslam, M., Bhobe, R., Alem, N., Donthu, S., & Dravid, V. P. (2005). Controlled large-scale synthesis and magnetic properties of single-crystal cobalt nanorods. *Journal of Applied Physics, 98*, 1–8.

203. Somorjai, G. A., & Blakely, D. W. (1975). Mechanism of catalysis of hydrocarbon reactions by platinum surfaces. *Nature, 258*, 580–583.

204. Yang, H. G., et al. (2008). Anatase TiO_2 single crystals with a large percentage of reactive facets. *Nature, 453*, 638–641.

205. Liu, G., et al. (2009). Enhanced photoactivity of oxygen-deficient anatase TiO_2 sheets with dominant 001 facets. *The Journal of Physical Chemistry C, 113*, 21784–21788.

206. Han, X., Kuang, Q., Jin, M., Xie, Z., & Zheng, L. (2009). Synthesis of titania nanosheets with a high percentage of exposed (001) facets and related photocatalytic properties. *Journal of the American Chemical Society, 131*, 3152–3153.

207. Xia, Y. N., Xiong, Y. J., Lim, B., & Skrabalak, S. E. (2009). Shape-controlled synthesis of metal nanocrystals: Simple chemistry meets complex physics? *Angewandte Chemie-International Edition, 48*, 60–103.

208. Xi, G. C., & Ye, J. H. (2010). Synthesis of bismuth vanadate nanoplates with exposed 001 facets and enhanced visible-light photocatalytic properties. *Chemical Communications, 46*, 1893–1895.

Chapter 2
Experimental Development

2.1 Reaction System

In order to perform photocatalytic tests on compounds, a suitable reactor was fabricated based on previous successful designs in the literature [1–4]. The photocatalytic water splitting reaction is normally monitored by recording amount of product formed in a relatively small fixed volume batch reactor, over a set period of time. Therefore it is suitable to use a simple gas tight batch reactor for lab based testing, with an optical window for illumination. Most reactors used in the field can either be externally illuminated by a light source from the top down, or using a horizontal window (internal illumination is not considered). In this study, a side window will be used, as the choice of stirring method causes a vortex, which drives water and semiconductor particle away from the centre of the reactor to the perimeter. Thus, photons from the top down will be incident on the centre of the reactor, where the concentration of the solution is low (Fig. 2.1).

In order to stir the semiconductor-electrolyte (e.g. photocatalyst-water) mixture, a magnetic stirrer was used to drive an inert PTFE magnetic stirring bar in the solution, as shown in Fig. 2.1. Conventional top down steel impeller blades would not be suitable for this application because of the risk of contamination of the blades with the semiconductor-electrolyte, i.e. through metallic leaching of nickel or chromium species. The nature of heterogeneous photocatalysis is such that the main interactions are that of the absorption of photons by the semiconductor, cleavage of water using charge carriers, and then release of product (H_2, O_2 or both). In order for the photocatalyst to homogeneously absorb light (ignoring reflection), the photocatalyst particles must be mixed at a high enough radial velocity so that they do not sink to the bottom of the reactor, and maintain in constant motion close to the reactor walls. Having apparatus such as baffles would not benefit the absorption of light by the semiconductor since excessive mixing with water will not alter the reaction significantly. Water is later oxidised or reduced instantaneously, and gaseous product released as a dissolved gas in water. Therefore in this case, using a PTFE magnetic stirring bar to produce a radial

© Springer International Publishing Switzerland 2015   55
D.J. Martin, *Investigation into High Efficiency Visible Light Photocatalysts
for Water Reduction and Oxidation*, Springer Theses,
DOI 10.1007/978-3-319-18488-3_2

(a)   **(b)**

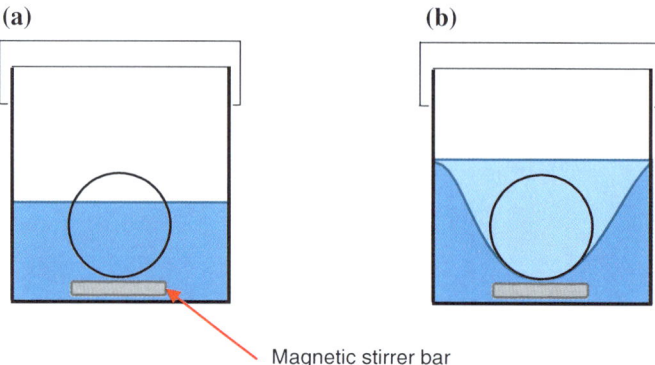

Magnetic stirrer bar

Fig. 2.1 Schematic of proposed reactor **a** without magnetic stirring, **b** with stirring

flow, and thus induce solid body rotation in the liquid—which is enough to form a temporary suspension—is more than satisfactory. More complex mixing is not necessary.

2.1.1 Reactor

A borosilicate cylindrical glass reactor was designed and then handmade by Labglass Ltd (4.5 cm radius, 11.5 cm height). The total volume was 730 cm^3, including headspace (calculated by water displacement). The vessel is fitted with a flat high purity borosilicate side window, which is slightly bigger than the beam of the light source (40 mm Φ reactor window, 33 mm Φ beam). High purity borosilicate is suitable for a photocatalytic reactions as the low impurity content (such as iron) allows both UV and visible wavelengths of light to pass through without any absorption (Fig. 2.2).

The reactor was also fitted with two side ports (fitted with GL18 aperture caps and silicon septa) for purging and gas sampling. The top of the system was a PTFE screw thread aperture cap, with a detachable transparent borosilicate window. The reactor sealing mechanism was originally a 7 mm depth silicon ring which fitted between the glass and the top PTFE aperture cap. However, it was discovered that the sealing provided by the company was insufficient; the sealing would become dislodged during purging, and also during experiments. Despite numerous replacements from LabGlass, the sealing was a continuous problem and frequently disrupted and forced experiments to be stopped. Ultimately a solution was found whereby a very thin piece of silicon (0.5 mm, Altec Ltd) was cut to fit the reactor diameter. This method of sealing provided not only an airtight seal throughout the experiment, but also was able to withstand higher pressures generated when a high purging flow rate (in order to accelerate purge time).

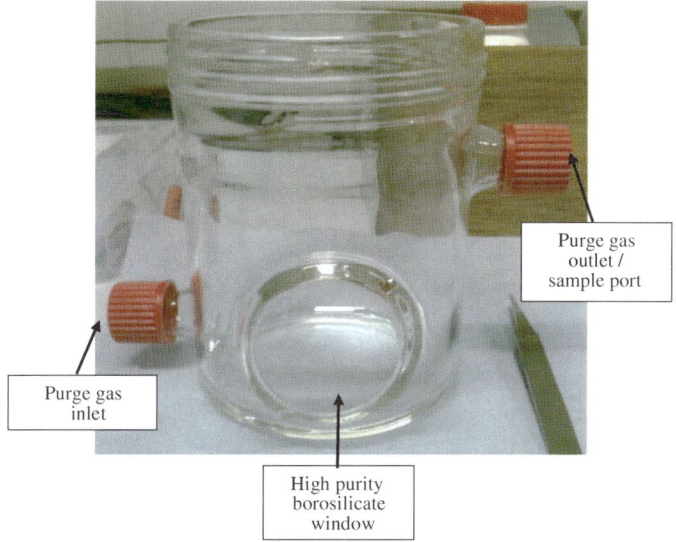

Fig. 2.2 Borosilicate reactor for water splitting batch reactions

2.1.2 Light Source

Two different xenon (Xe) lamps were purchased from Newport Spectra and TrusTech to act as artificial light sources. Xe lamps have a spectral profile akin to that of the solar spectrum, with intensity less than 20 mW m^{-2} nm^{-1} at $250 < \lambda < 800$ nm[15]. A 300 W Xe lamp (TrusTech PLS-SXE 300/300UV) was used for oxygen, hydrogen, and water splitting, and also for calculating solar-to-hydrogen conversion efficiency. The higher power lamp source will in theory enable the probing of photocatalysts whose efficiencies are considerably small, as a larger photon flux will increase the rate of water splitting. A 150 W Xe lamp (Newport 6256 150 W Xe) was utilised for IQY measurements; the lower intensity of the 150 W Xe lamp prevented damage to the band pass filters, which absorb considerable amounts of light, up to 90 %.

The power from the lamp was calculated using a Silicon photodiode detector (190–1110 nm), with built in attenuator, connected to a handheld digital power meter (both purchased from Newport Spectra). Various long pass filters were used, from 400 to 550 nm, supplied by Comar Optics; enabling the selective use of either full arc or restricted visible light. Similarly, band pass filters were used in IQY measurements. These filters have a quoted centre wavelength, for example 400 nm, and the width of all pass-bands are 10 nm (395–405 nm).

2.2 Gas Chromatography: Selection and Calibration

2.2.1 Gas Chromatography Setup

A Varian 450 gas chromatograph (GC) was used to analyse the amounts of gaseous products from water splitting reactions (Fig. 2.3), and to monitor nitrogen levels within a batch reactor. Samples were taken by using a gas-tight syringe (Hamilton® 1000 µL). The GC was fitted with a TCD and molecular sieve 5A column, running with argon carrier gas (zero grade). As discussed in Sect. 1.2.2, argon is a suitable carrier gas for analysing concentrations of hydrogen, oxygen and nitrogen, due to the sufficiently low thermal conductivity—therefore all produced peaks will be positive, making integration easier during analysis (Table 2.1). Helium is an alternative, but is slightly more expensive, and would also yield negative peaks for oxygen and nitrogen due to the higher thermal conductivity, which would then be problematic during peak analysis. For these reasons it was not selected.

The signal from the GC's TCD produces a chromatogram on the Varian 450-GC software (an idealised version is shown in Fig. 2.4). The chromatograph is a plot of signal (µV) versus time (minutes); whereby the area under the signal

Fig. 2.3 Photograph of the gas chromatograph unit used during photocatalysis experiments

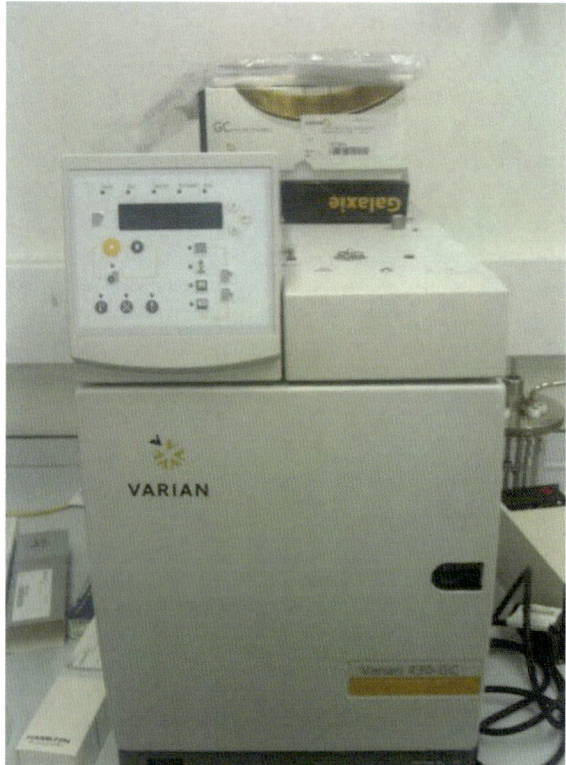

Table 2.1 Gases and corresponding thermal conductivity at STP [5]

Gas	Thermal conductivity ($W\,m^{-1}\,K^{-1}$)
Hydrogen	0.1805
Helium	0.1513
Nitrogen	0.0259
Oxygen	0.0266
Argon	0.0177

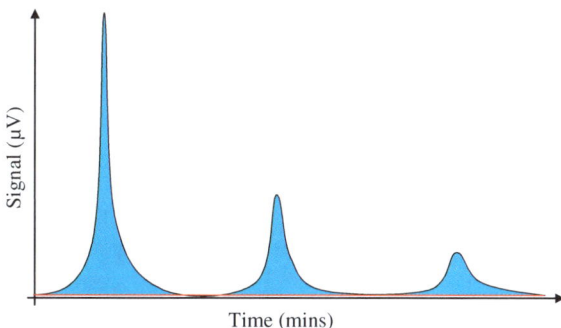

Fig. 2.4 Idealised example chromatogram. A baseline is indicated by a horizontal *red* line, and integrated area under the curve (in μV min) is indicated by a shaded *blue* area

(μV min) is proportional to the concentration of a specific gas, and each separate signal is separated in time, denoting different gases. By selecting an appropriate baseline, and performing an automatic integration via the software, an accurate signal area can be acquired.

In order to establish a reasonable signal, parameters were tuned to optimise peak separation, signal-to-noise ratio, and baseline fluctuations. Many water splitting experiments documented in the literature record gas concentrations at either 15 or 30 min intervals, with some elongated experiments recording every hour [3]. Therefore the total run time for each chromatogram should be less than 15 min, i.e. all necessary gases should appear in a 10 to 15 min window, or have a retention time (RT) less than 15 min. In gas chromatography, the parameters which governs RT of a gas are oven temperature, pressure and flow rate of carrier gas inside the column—two of which were consequently set to 50 °C, and 50 PSI to enable all gases (hydrogen, oxygen, nitrogen) to come through within a 10 min window. Signal-to-noise ratio and baseline fluctuations can be controlled by adjusting flow rate and TCD filament temperature. After optimisation, the most efficient flow rate was 10 cm^3 min^{-1} and 180 °C. In general, increasing the flow rate through the TCD decreased peak width, but dramatically decreased peak height/response. Therefore a low flow rate was chosen so that small concentrations of gases could be detected.

2.2.2 Standard Gas and Calibration

In order to correlate the response generated by the TCD on the GC to the actual gas composition, a calibration curve is used. By injecting known concentrations (and thus known molar amounts) of gas, and then monitoring the response, a graph of area (y) versus concentration (x) can be plotted. Then, by solving the equation of a line (y = mx + c), for any area, a concentration can be calculated.

To then acquire an accurate known concentration, a suitable standard gas was purchased from BOC. It comprises of 99 % zero grade argon, with appropriate amounts of H_2 (4000 ppm) and O_2 (2000 ppm), among other gases which might be produced as by-products in a half reaction in the presence of charge scavenger (1000 ppm CO, 1000 ppm CO_2 and 2000 ppm methane). By injecting different volumes of this gas into the GC, a calibration curve can be built (Fig. 2.5), along with a response factor (R). The R factor is calculated by plotting amount of gas versus area, then solving the equation of the straight line for where the line intercepts the x-axis. Generally the response factor (essentially an offset error) should be as close to zero as possible, implying a true linear relationship between gas amount and peak area. In reality however, this is not the case, due to errors in manual syringe sampling, TCD response and for oxygen error—a residual amount of air left in the syringe (dead volume). However, these can be corrected for simply using the R-factor in calculating the amount of gas being measured.

To establish a sampling error, 10 different samples of 0.5 cm^3 standard gas was injected, and then average, standard deviation (SD, σ) and percentage error was calculated (Table 2.2). The percentage error was calculated by dividing the SD by the mean.

Using Table 2.2, it is possible to apply a percentage error on both future H_2 and O_2 sampling data, which can be applied to calibration curves (Fig. 2.5, Table 2.3).

According to the linear fit statistics, an unknown molar amount of hydrogen or oxygen can be calculated by knowing the area under the curve from a gas sample. For example an area of 86,000 μV min is recorded by the GC for hydrogen. Using the equation of the line, y = 55663x + 216.54;

$$\left(\frac{86000 - 216.54}{55663}\right) = 1.54\ \mu mol\ of\ hydrogen \qquad (2.1)$$

The linear regression best fit line shows that for hydrogen calibration, $r^2 = 0.9994$, and for oxygen $r^2 = 0.9947$. These high values of r^2 also confirm that by knowing y, predicting x using y = mx + c has a high degree of confidence/accuracy [6]. Essentially r-squared is a fraction which is used to determine how x changes linearly with y.

It is important to separate the oxygen measured from the standard gas, and that from air, which is left in the syringe tip (dead volume). From Table 2.2 it is evident that even a small amount of air in the syringe tip can influence the calibration,

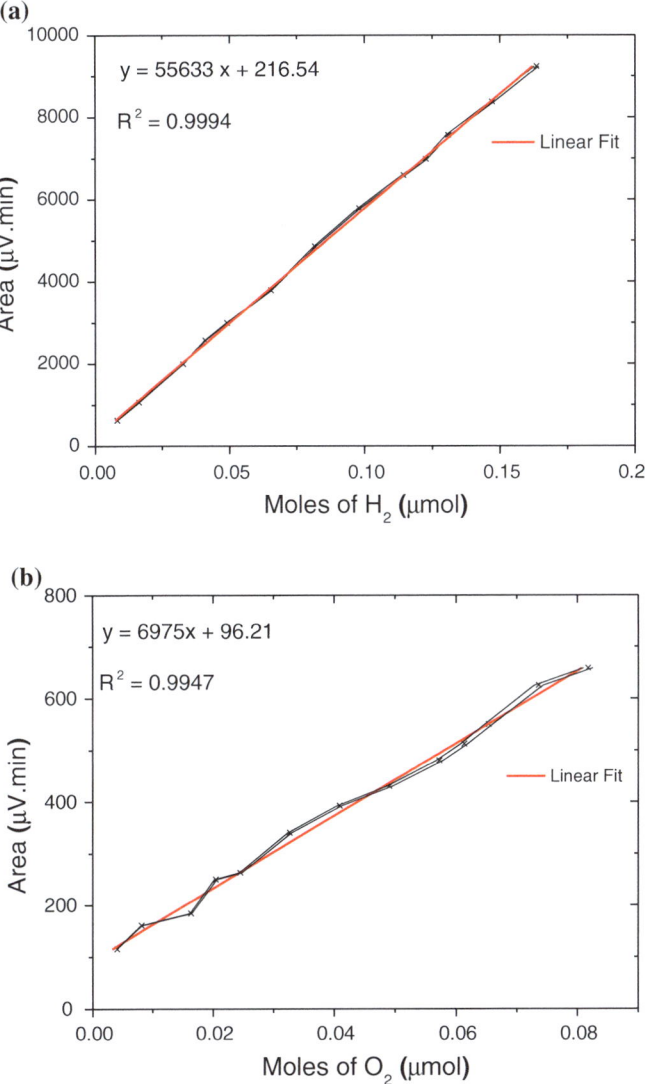

Fig. 2.5 GC area versus molar amount calibration curves for hydrogen (**a**), and oxygen (**b**). Equations of linear fits and R^2 values are noted in the *upper left* hand corner of the figure

increasing error—which is nearly double that of the standard error on a hydrogen sample, since no hydrogen exists freely in air. The air contaminant also comes from a small amount which is inside the injection port in the GC. All of which is taken into account for the calibration. The successful calibration of the GC means that gas amounts as low as 10^{-9} mol can be detected accurately.

Table 2.2 Area sampling data for H_2 and O_2 (0.5 cm^3)

	Area (μV min)	
	H_2	O_2
	4896.1	390.5
	4834.4	394.5
	4831.0	389.2
	4885.1	388.2
	4876.7	399.5
	4867.5	397.6
	4845.9	391.1
	4849.8	393.3
	4888.6	393.9
	4820.3	399.0
Mean	**4859.5**	**393.7**
σ	25.4	3.8
% Error	0.52 %	0.97 %

SD was taken for entire population

2.3 General Characterisation

The following section details standard characterisation methods that are used for all necessary results sections, to avoid repetition.

2.3.1 UV-Vis Spectrophotometry

Absorption, reflection and transmission spectra were collected from a Shimadzu UV-2550 spectrophotometer fitted with an integrating sphere. The software provided (UV-Probe 2.33) enabled reflection to be directly converted to absorption by the Kubelka-Munk transformation. Typically, data would be collected from 250 to 800 nm, with an optimum slit with of 2 nm. This reduces noise whilst not compromising the accuracy of the data. The frequency of the data was 0.5 nm, as this was more than adequate enough to determine precise spectra.

2.3.2 PXRD

PXRD was performed on either a Rigaku RINT 2100 (40 kV, 40 mA, using a Cu source with $K_{a1} = 1.540562$ and $K_{a2} = 1.544398$) or a Bruker D4 (40 kV, 30 mA, using a Cu source with $K_{a1} = 1.54056$ and $K_{a2} = 1.54439$). A maximum 0.05° step size was used, at 5 s per step, covering a maximum range of 0–90° (2θ). Phase match and baseline corrections were performed on either MDI Jade, or

Table 2.3   Data table for Fig. 2.5

Hydrogen

Syringe volume (cm^3)	Amount of H$_2$ (cm^3)	Amount of H$_2 \times 10^{-9}$(m^3)	Moles of H$_2 \times 10^{-8}$	Moles of H$_2$ (μM)	Area (μV min)
0.05	0.0002	0.2	0.82	0.008	623.4
0.1	0.0004	0.4	1.64	0.016	1059.5
0.2	0.0008	0.8	3.27	0.033	1992.3
0.25	0.0010	1.0	4.09	0.041	2575.9
0.3	0.0012	1.2	4.91	0.049	3000.5
0.4	0.0016	1.6	6.54	0.065	3787.9
0.5	0.0020	2.0	8.18	0.082	4853.8
0.6	0.0024	2.4	9.82	0.098	5784.0
0.7	0.0028	2.8	11.50	0.115	6590.5
0.75	0.0030	3.0	12.30	0.123	6982.6
0.8	0.0032	3.2	13.10	0.131	7565.7
0.9	0.0036	3.6	14.70	0.147	8372.0
1	0.004	4.0	16.30	0.164	9237.7

Oxygen

Syringe volume (cm^3)	Amount of O$_2$ (cm^3)	Amount of O$_2 \times 10^{-9}$(m^3)	Moles of O$_2 \times 10^{-8}$	Moles of O$_2$ (μM)	Area (μV min)
0.05	0.0001	0.1	0.41	0.0041	115.5
0.1	0.0002	0.2	0.82	0.0082	160.7
0.2	0.0004	0.4	1.64	0.0164	184.1
0.25	0.0005	0.5	2.04	0.0204	250.0
0.3	0.0006	0.6	2.45	0.0245	263.1
0.4	0.0008	0.8	3.27	0.0327	340.0
0.5	0.0010	1.0	4.09	0.0409	392.4
0.6	0.0012	1.2	4.91	0.0491	430.0
0.7	0.0014	1.4	5.73	0.0573	480.1
0.75	0.0015	1.5	6.13	0.0613	512.0
0.8	0.0016	1.6	6.54	0.0654	550.3
0.9	0.0018	1.8	7.36	0.0736	625.1
1	0.0020	2.0	8.18	0.0818	658.4

Bruker's EVA software using the ICSD/JCPDS database. A powdered sample was flattened into an amorphous glass (Rigaku) or silicon (Bruker) well holder.

2.3.3   FE-SEM

A JEOL JSM-7401F was used for measuring particle size, examining agglomeration of particles and also performing EDX measurements. The main advantage of using a Field Emission-SEM is that a greater resolution is possible (up to 6 times greater than conventional SEMs). Furthermore, electrostatic charging is greatly reduced on poorly conductive samples because lower acceleration voltages, without compromising image quality. Carbon tape was used as a conductive adhesive for the powdered samples.

2.3.4 TEM

Conventional TEM measurements were taken using a JEOL2010F, at an accelerating voltage of 200 keV. Powdered samples were diluted in chloroform, sonicated to disperse particle, and then dropwise added onto a conductive copper grid. Tilt studies and EDX measurements were also conducted on a JEOL2010F.

2.3.5 BET Specific Surface Area

Specific surface area was calculated via the BET method, using N_2 absorption by a Micromeritics TriStar 3000. Powdered samples were placed in a borosilicate vial, with a weight to surface area ratio of 1 g to 30 m^2 in order to acquire suitable data. Values of R^2 (linear regression) were tuned to be as close to 1 as possible, with all values ≥ 0.9998.

2.3.6 ATR-FTIR Spectroscopy

ATR-FTIR spectroscopy was performed on a Perkin-Elmer 1605 FT-IR spectrometer in the wavenumber range from 400–4000 cm^{-1} with a resolution of 0.5 cm^{-1}. Powdered samples were placed on the ATR crystal, and then compressed using a flat axial screw. Spectra were compared with literature examples, as there was available universal 'search and match' library for the instrument used.

2.3.7 Raman Spectroscopy

Raman spectroscopic measurements were performed on a Renishaw InVia Raman Microscope, using an Ar^+ 514.5 nm excitation laser, and a wavenumber range from 100–2000 cm^{-1}. A 'notch filter' was used to cut out Raleigh components, and a silicon standard was used for calibration at 520 cm^{-1}.

2.3.8 TGA-DSC-MS

Thermo Gravimetric Analysis-Differential Scanning Calorimetry-Mass Spectroscopy (TGA-DSC-MS) was performed on a Netzsch Jupiter TGA-DSC, connected to a Netzsch Aeolius MS, in an inert He atmosphere. The data was then processed using the Netzsch 'Proteus' thermal analysis software. Precursors (urea, thiourea) were placed in an alumina crucible, and calcined from 26.9 to 600 °C, over a period of 115 min. From the raw data, TGA data yields a plot of mass loss (%) versus temperature, DSC yields heat change (exothermic/endothermic reaction) versus temperature information, and MS monitors mass number (ion current, nA) relative concentration versus temperature.

2.3.9 Zeta Potential (ZP) Measurements

ZP measurements were performed using a Zetasizer nano ZS equipped with a He–Ne laser 633 nm, maximum 4mW power, and analysis was undertaken using the 'Zetasizer software'. Powdered samples were diluted in a 0.05 M aqueous NaCl solution, which acted as an electrolyte, increasing conductivity. A 5 cm^3 sample was then sonicated for 30 min, and then placed into a ZP cuvette/cell with gold connectors. ZP was monitored as pH was changed from neutral (starting at ca. 7.7) to both alkaline and acidic conditions using NaOH and HCl respectively (manual titration). For each pH point, 5 measurements were taken, and the mean then calculated. The isoelectric point (IEP) is defined as the point of zero change at a set pH, and thus was recorded at ZP $= 0$.

2.3.10 XPS

XPS measurements were performed on a Thermoscientific XPS K-alpha surface analysis machine using an Al source. Analysis was performed on the Thermo Advantage software. After samples were placed under UHV, a sweep scan was performed from 100–4000 eV. Each sample was scanned 6 times at different points on the surface to eliminate point error and create an average. Specific elemental peaks were then identified, and analysed further.

2.3.11 Elemental Analysis

Elemental Analysis (EA) was performed on a Micro Elemental Analyzer (CE-400 CHN Analyser, Exeter Analytical Instruments). Accurate (± 0.1 %) weight percentages of carbon, nitrogen, hydrogen and trace elements were converted to atomic percentages before analysis.

References

1. Yi, Z., et al. (2010). An orthophosphate semiconductor with photooxidation properties under visible-light irradiation. *Nature Materials, 9*, 559–564.
2. Murphy, A. B., et al. (2006). Efficiency of solar water splitting using semiconductor electrodes. *International Journal of Hydrogen Energy, 31*, 1999–2017.
3. Chen, X., Shen, S., Guo, L., & Mao, S. S. (2010). Semiconductor-based photocatalytic hydrogen generation. *Chemical Reviews, 110*, 6503–6570.
4. Cabrera, M. I., Alfano, O. M., & Cassano, A. E. (1994). Novel Reactor for Photocatalytic Kinetic Studies. *Industrial and Engineering Chemistry Research, 33*, 3031–3042.
5. Tsederberg, N. V., & Cess, R. D. (1965). *Thermal conductivity of gases and liquids*. Cambridge, Massachusetts: MIT Press.
6. Seber, G. A. & Lee, A. J (2012). *Linear regression analysis* (Vol. 936). New York:Wiley.

Chapter 3
Oxygen Evolving Photocatalyst Development

This chapter details the development and understanding of the extremely efficient photocatalyst, Ag_3PO_4. Due to the lack of literature which followed on from the initial investigation (by Yi et al.) exploring the photooxidation of water, it was envisaged that the preparation method was difficult to replicate [1]. This was indeed the case, as results show there was considerable differences in oxygen evolution from water between samples synthesised using different phosphate precursor sources (PO_4^-), and different solvents. A method was established to fabricate a silver phosphate sample which had a similar roughly spherical morphology, and exhibited a photooxidative ability analogous to that published, using ethanol, $AgNO_3$, and Na_2HPO_4. Most notably, when the phosphate precursor was changed to H_3PO_4, and using a large volume of ethanol was used, it was possible to slow the growth of Ag_3PO_4 crystals, and retain low index facets, resulting in the production of novel {111} terminated tetrahedral Ag_3PO_4 crystals. By changing the concentration of H_3PO_4, and thus changing the reaction kinetics, it is possible to monitor the growth mechanism of Ag_3PO_4 crystals from tetrapods to tetrahedrons.

Ag_3PO_4 tetrahedrons demonstrate considerably higher oxygen evolution activity in comparison to roughly spherical particles, and other low index facets. DFT calculations were used to model the surface energy, and the hole mass of Ag_3PO_4 in different directions. It is concluded that a combination of high surface energy and low hole mass on Ag_3PO_4 {111} surfaces result in extremely high oxygen evolution from water, with an internal quantum yield of 98 % [2].

3.1 Introduction

Water photolysis for H_2 fuel synthesis has the potential to solve both increasing demand of renewable energy and climate change caused by CO_2 emissions. However the search for a solitary low cost semiconductor which has a band gap suitable for an efficient, neutral water splitting reaction under ambient conditions has been met with little success. Nature, however, demonstrates an

© Springer International Publishing Switzerland 2015
D.J. Martin, *Investigation into High Efficiency Visible Light Photocatalysts for Water Reduction and Oxidation*, Springer Theses,
DOI 10.1007/978-3-319-18488-3_3

efficient strategy to utilise solar irradiation (nearly unity IQY) by spatially and temporally separating electrons and holes in wireless photosynthesis reactions [3]. So, water splitting can be envisaged as two half reactions; water oxidation, and an equivalent of proton reduction to hydrogen fuel. The former of the two is much more challenging because one molecule of gaseous oxygen requires 4 holes, and occurs on a timescale ca. 5 orders of magnitude slower than H_2 evolution, proven in both natural and artificial photosynthesis [4–6]. Therefore, finding an relatively cheap, robust and efficient water oxidation photocatalyst is widely accepted to be key to solar driven fuel synthesis, if it is to be commercial viable.

Ag$_3$PO$_4$ is used commonly in the pharmaceutical field as an antibacterial agent [7], and is also considered to be an ideal candidate for water photooxidation due to an appropriate band gap position, non-toxicity, and high photocatalytic ability [1, 8]. It is widely known that activity of a catalyst can be determined by its exposing facets. In order to attain specifically terminated facets, morphology controlling agents (MCAs), such as organic surfactants (e.g. PVP), or capping agents (e.g. fluorine ions [9]) are commonly employed. Due to strong interaction with the substrate, complete removal of these MCAs is very challenging, but absolutely essential for effective catalytic activity in water splitting. One of the major research obstacles facing researchers today is developing a facile, MCA-free synthesis route to attain clean, reactive facets, for example, by kinetic control [10].

We performed DFT calculations, and it was found that Ag$_3$PO$_4$ possessed an anisotropic hole mass (m_h^+), and variable surface energy depending on the exposing facets—in particular that the {111} crystal planes had a large surface energy. Previously, photocatalytic organic decomposition on faceted Ag$_3$PO$_4$ has been reported. For example, Wang and co-workers synthesised tetrahedral shaped particles after attempting to rejuvenate Ag-Ag$_3$PO$_4$. Due to the presence of silver metal within the Ag$_3$PO$_4$ particles, they exhibited lower activity for the degradation of the dye RhB in comparison to mixed faceted particles [11]. Bi and co-workers have also compared photocatalytic degradation reactions over cubic and rhombic dodecahedral silver phosphate structures [12]. However, to date, there have not been reports on the influence of different exposing Ag$_3$PO$_4$ facets on water photooxidation, which is a well-known rate determining step in photocatalytic water splitting and has different requirements in comparison to organic decomposition reactions. Therefore, attempts were made to manipulate the material facets in order to control both charge transport (in particular hole transport and mobility) and surface photooxidation reactions using a facile and environmentally friendly approach. Herein, this chapter shows the manipulation of Ag$_3$PO$_4$ facets, and for the first time investigates their individual ability for water photooxidation and details the underlying mechanisms which control oxygen production; specifically between different low index facets of Ag$_3$PO$_4$ using diverse characterisation methods.

3.2 Methodology

3.2.1 Photocatalytic Analysis

Photooxidation reactions were carried out in both the UK and Japan to verify the continuity of results, which were deemed to be consistent. For a typical oxygen evolution experiment, 0.2 g of Ag_3PO_4 powder was dispersed in 230 cm^3 of deionised water, with an additional amount of 0.85 g $AgNO_3$ acting as an electron acceptor. $AgNO_3$ was used as an electron scavenger in all standard oxygen evolution experiments as it was shown by Mills et al. to be the most efficient, and convenient electron scavenger [13]. The whole reaction was carried out in a custom glass reactor. Before the solution was irradiated, the reactor was either connected to a vacuum system (Japan), and evacuated, or thoroughly purged with argon (UK) to remove all oxygen in the headspace of the reactor and dissolved oxygen in water. A baseline was taken to ensure that there was little or no detectable oxygen in the system. A 300 W Xe lamp (UK: TrusTech PLS-SXE 300/300UV; Japan: Hayashi Tokei, Luminar Ace 210,) was used to irradiate the sample after a suitable baseline was obtained (power density full arc: ca. 500 ± 25 mW cm^{-2}, 420 nm filter: ca. 360 ± 20 mW cm^{-2}). A small sample of gas was sent to a gas chromatograph (UK: Varian 430-GC, TCD, Argon carrier gas, Japan: Shimadzu GC-8A, TCD, Argon carrier gas,) at specific intervals, upon which the concentration of oxygen was identified.

Internal quantum yield measurements were measured in the UK by inserting a band pass filter (365, 400, 420, 500, 600 nm) in front of a 150 W Xe light source (Newport Spectra) to obtain the correct wavelength. The light intensity was measured at 5 different points and then and average of the light intensity was then taken. The efficiency was then calculated by the standard formula quoted in literature, ignoring reflected photons [14–16]; IQY(%) $= 4 \times$ number of O_2 molecules detected divided by the number of photons absorbed by the photocatalyst. The reported turnover frequency (TOF min^{-1}) was calculated using the following formula as quoted in the literature [14, 15]: TOF (min^{-1}) $=$ the ratio of number of reacted holes to number of atoms in photocatalyst, per minute over a 3 h experiment. Turnover number was calculated using the same formula, negating time.

3.2.2 Synthesis Techniques

Ag_3PO_4 was originally fabricated as an active photocatalyst by Yi et al. using an ion-exchange method [1]. The authors claim that by mixing/grinding stoichiometric amounts of either Na_2HPO_4 (anhydrous, 'dibasic') or $Na_3PO_4 \cdot 12H_2O$ (dodecahydrate, 'tribasic') with $AgNO_3$, at room temperature using no solvents, a Ag_3PO_4 precipitate is formed. This precipitate is then washed with deionised water to remove unreacted precursors and dried at 70 °C to remove all water. This method (denoted 'A') was repeated with both precursors (99.9 % grade, purchased from Sigma-Aldrich and VWR), using a pestle and mortar for mixing the two solids. NaH_2PO_4 (anhydrous, 'monobasic', 99.9 %, VWR) was also mixed with $AgNO_3$ in a similar manner for completion. Method A and subsequent methods B-D are summarised in Table 3.1.

This method was repeated with four phosphate precursors (Na_3PO_4, Na_2HPO_4, NaH_2PO_4, H_3PO_4), but with the addition of 5 cm^3 of ethanol or methanol (95 %, Merck Schuchardt) to aid ion-exchange during the mortar and pestle grinding process (denoted method 'B'). No powder sample was synthesised when H_3PO_4 was used, as the acidity of phosphoric acid (pH 1) prevented the formation of Ag_3PO_4.

In a different method (denoted 'C'), stoichiometric amounts of Na_2HPO_4 and $AgNO_3$ were separately dissolved in 80 cm^3 of ethanol in a conical flask, with the aid of a magnetic stirrer and magnetic stirring plate (approx. 1500 rpm). Then the $AgNO_3$-ethanol solution was added dropwise to the phosphate-ethanol solution until a yellow Ag_3PO_4 precipitate was formed. The solution was both centrifuged and washed (3 times) immediately after precipitation, or after 1 h of stirring.

Using method 'C' as a template, Na_2HPO_4 was replaced with another phosphate precursor which was free of sodium, H_3PO_4 (85 %, Acros Organics). This method ('D') involved the same initial steps in terms of dissolving both liquid H_3PO_4 (various volumes from 2 to 20 cm^3) and 2 g $AgNO_3$ in separate ethanol solutions of 80 cm^3 volume under vigorous stirring. Upon adding the $AgNO_3$-ethanol solution to the H_3PO_4-ethanol solution, a yellow Ag_3PO_4 precipitate appeared, but then rapidly dissolved, most likely due to the pH of the H_3PO_4-ethanol solution (approx. pH 1), as seen in method 'B'. After adding

Table 3.1 Summary of results on initial Ag_3PO_4 studies using various phosphate precursors

Method	Sample name	Phosphate precursor	Synthesis method	Initial oxygen evolution rate (μmol h^{-1} g^{-1}) under visible light ($\lambda \geq 420$ nm). Error = 0.97 %	BET SSA (m^2 g^{-1})
A	A1	Na_2HPO_4	Ground, No solvent	733	1.12
A	A2	$Na_3PO_4 \cdot 12H_2O$		478	1.35
A	A3	NaH_2PO_4		687	0.98
B	B1	Na_2HPO_4	Ground, 5 cm^3 EtOH	1357	1.01
B	B2	NaH_2PO_4		671	1.40
B	B3	Na_2HPO_4	Ground, MeOH	1159	1.29
C	C1	Na_2HPO_4	Dissolved in EtOH	1418	0.91
D	Tetrahedron	20 cm^3 H_3PO_4	Dissolved in EtOH	6074	1.01

All Ag_3PO_4 compounds were synthesised using $AgNO_3$ as the silver ion source approximately 10 cm^3 of AgNO$_3$-ethanol solution, the H$_3$PO$_4$-ethanol solution turned slightly milk-white. If all 80 cm^3 of the AgNO$_3$-ethanol solution was added, the colour remained milk white; no precipitate was formed. If however, 10 cm^3 of AgNO$_3$-ethanol was added to the H$_3$PO$_4$-ethanol solution, and then a few drops of the solution added back into the bare AgNO$_3$-ethanol solution, an instantaneous yellow precipitate was formed. After 1 h of stirring the yellow precipitate turned dark green. The precipitates were centrifuged and washed with deionised water 3 times, before drying overnight in an oven at 70 °C.

Rhombic dodecahedrons and cubic Ag$_3$PO$_4$ crystals were synthesised using a recipe based on a study by Bi et al. [12]. The initial recipe described by the authors did not yield the required morphology, and therefore a different method was developed.

To synthesise Ag$_3$PO$_4$ cubes, 0.45 g AgNO$_3$ was dissolved in 50 cm^3 DI water. Meanwhile in a separate vial, 0.069 cm^3 NH$_3$·H$_2$O was added dropwise to 10 cm^3 DI water. The ammonia solution was then added to the silver nitrate solution, to which the resulting solution turned first brown (Ag$_2$O), and then transparent [Ag(NH$_3$)$_2$]$^+$. This colour change can be explained by the pH mediated complexing of Ag$^+$ ions:

Initially, NH$_4$OH dissociates and releases OH$^-$ ions

$$NH_3 \cdot H_2O \rightarrow NH_4OH \rightarrow NH_4^+ + OH^- \tag{3.1}$$

The OH$^-$ ions then react with silver ions (dissolved in the solution from AgNO$_3$)

$$2Ag^+ + 2OH^- \rightarrow Ag_2O + H_2O \tag{3.2}$$

And finally as the pH becomes more basic with the addition of more NH$_4$OH

$$Ag_2O + 4NH_3 + H_2O \rightarrow 2\left[Ag(NH_3)_2\right]^+ + 2OH^- \tag{3.3}$$

To this transparent solution, 0.15 M of aqueous Na$_2$HPO$_4$ solution was added until the solution appeared yellow-green. The solution was stirred for 2 h at 1000 rpm using a magnetic stirrer-bar/plate, and subsequently washed and centrifuged 3 times to remove any unreacted product.

Bi et al. used PVP to synthesise rhombic dodecahedrons, however, when this method was repeated, the desired morphology was not seen. To synthesise silver phosphate rhombic dodecahedrons, in the absence of the surfactant PVP, the following novel recipe was implemented. 0.6 g AgNO$_3$ was dissolved in 300 cm^3 DI water. A 0.1 M ammonium hydroxide solution was added until the silver nitrate solution turned brown, and then clear. This signifies the formation of a silver hydroxide species, and then a silver ammonia complex respectively. A 0.15 M Na$_2$HPO$_4$ (450 cm^3) was then added to the silver ammonia complex solution. In order to neutralise the pH, H$_3$PO$_4$ was added dropwise, to which a bright yellow compound precipitated. The solution was also stirred for 2 h at 1000 rpm using a magnetic stirrer-bar/plate, and subsequently washed and centrifuged 3 times to remove any unreacted product.

3.3 Results and Discussion

3.3.1 Initial Ag₃PO₄ Studies

In order to establish a definitive baseline, three different synthesis methods (A–C) were attempted in order to acquire a sample with similar activity (1272 μmol h^{-1} g^{-1}) reported by Yi et al. under visible light (l > 420 nm). The results are summarised in Table 3.1.

3.3.1.1 Method 'A'

The three samples fabricated using a solid mixing/grinding method ('A') in a mortar and pestle all yielded relatively large agglomerates of smaller, roughly spherical particles, as can be seen in Fig. 3.1. All samples were confirmed my XRD to be phase pure body-centred cubic crystalline Ag$_3$PO$_4$ (6.004 Å, BCC, P-43n, JCPDS no.06-0505) (Fig. 3.2a), and possess an absorption profile characteristic of a semiconductor with a band gap to absorb light in the visible range (Fig. 3.2b). Particle size analysis was not performed as the difference between an agglomerate and a distinct particle could not be verified (Fig. 3.3).

Since there was no correlation between oxygen evolution rate (OER) and BET SSA, it was theorised that either unreacted sodium or nitrate ions in the solution and at the surface could be hindering activity, by adsorbing on the surface of the photocatalyst and blocking active sites. Therefore, during the solid mixing process, a small amount of ethanol/methanol was added to facilitate better ion exchange and potentially remove Na ions from the surface

3.3.1.2 Method 'B'

Since sample A2, synthesised using Na$_3$PO$_4$·12H$_2$O, displayed a poor OER, even with the largest surface area from the batch, it was decided to not use Na$_3$PO$_4$·12H$_2$O in the next method ('B'). It is speculated that either less Na in the precursor, or the presence of water in the synthesis method could influence surface adsorption of Na ions. Therefore EtOH and MeOH were used to facilitate ion exchange, and precursors without water were used (anhydrous NaH$_2$PO$_4$ and Na$_2$HPO$_4$).

As shown in Fig. 3.4, all B-series samples shown some degree of coagulation/agglomeration, with hugely irregular 'particle' size. With such an obviously large particle size range, it would be impossible to attribute an activity change to a definitive particle size effect. The XRD data (Fig. 3.5) shows that B1 is phase pure, whilst B2 and B3, have unknown sub-phases, which could not be identified. The UV-Vis absorbance spectra (Fig. 3.6) of the samples show all three samples exhibit an absorption edge typical of silver phosphate. Since there are no other

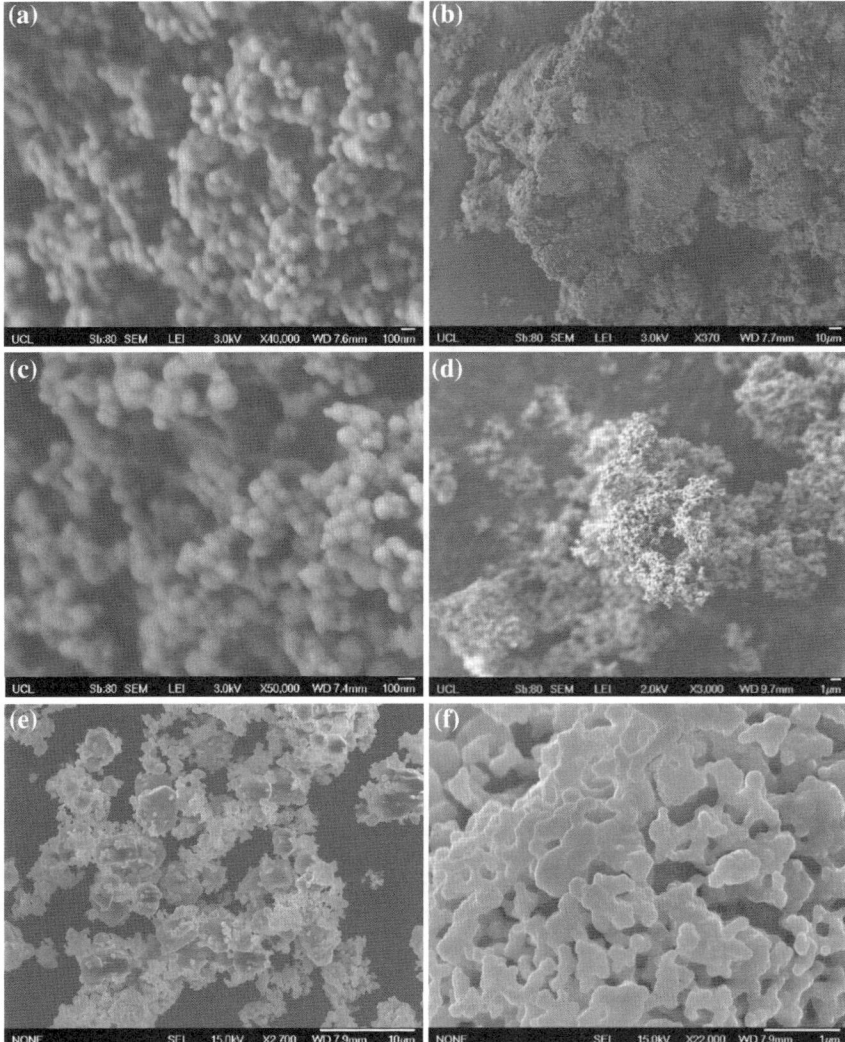

Fig. 3.1 SEM micrographs of samples A1 (**a**, **b**), A2 (**c**, **d**), and A3 (**e**, **f**)

detectable absorption band edges besides Ag_3PO_4, it is theorised that the unknown compound is not a semiconductor and therefore does not absorb light in an interband transition process. All three samples display similar BET SSA (Table 3.1), yet display OERs nearly double that of the samples synthesised using method 'A'—with the exception of B2 (Fig. 3.7). The unknown sub-phase produced during synthesis could be a result of the precursor, however, the presence of the sub-phase is certainly detrimental to the activity.

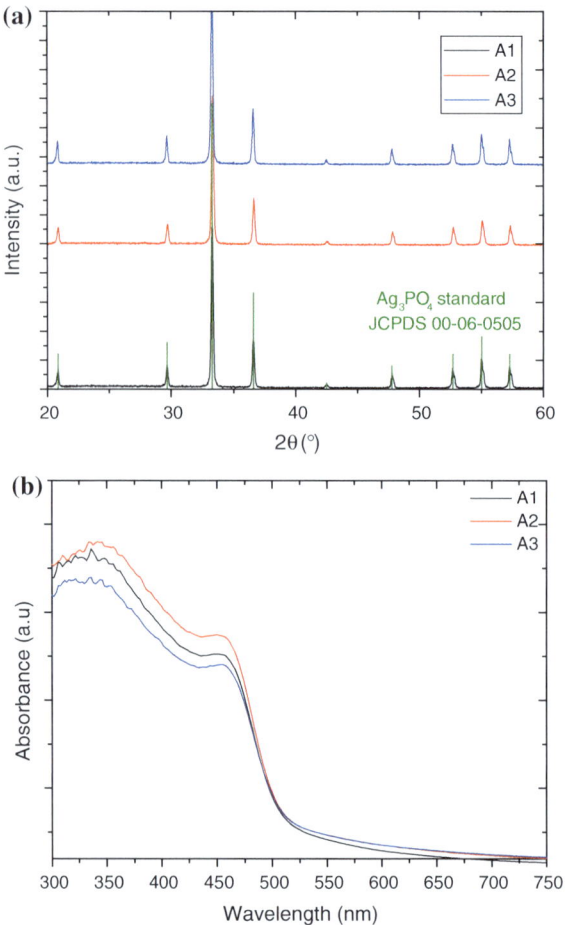

Fig. 3.2 **a** XRD patterns of samples *A1*, *A2* and *A3*. **a** UV-Vis absorbance spectra of samples *A1*, *A2* and *A3*

Fig. 3.3 Oxygen evolution under visible light ($l > 420$ nm) from water, in the presence of AgNO₃ (0.85 g), at neutral pH, using 0.2 g photocatalyst (*A1*, *A2*, *A3*)

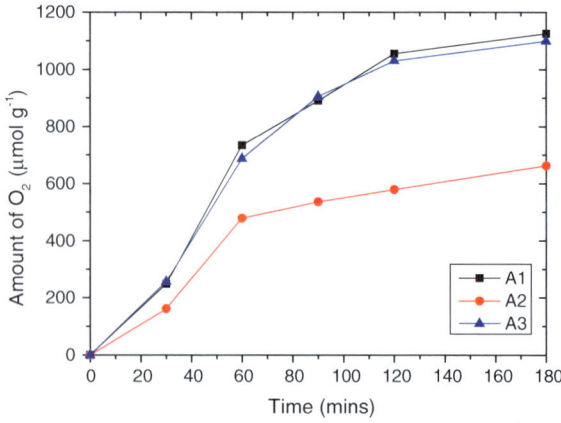

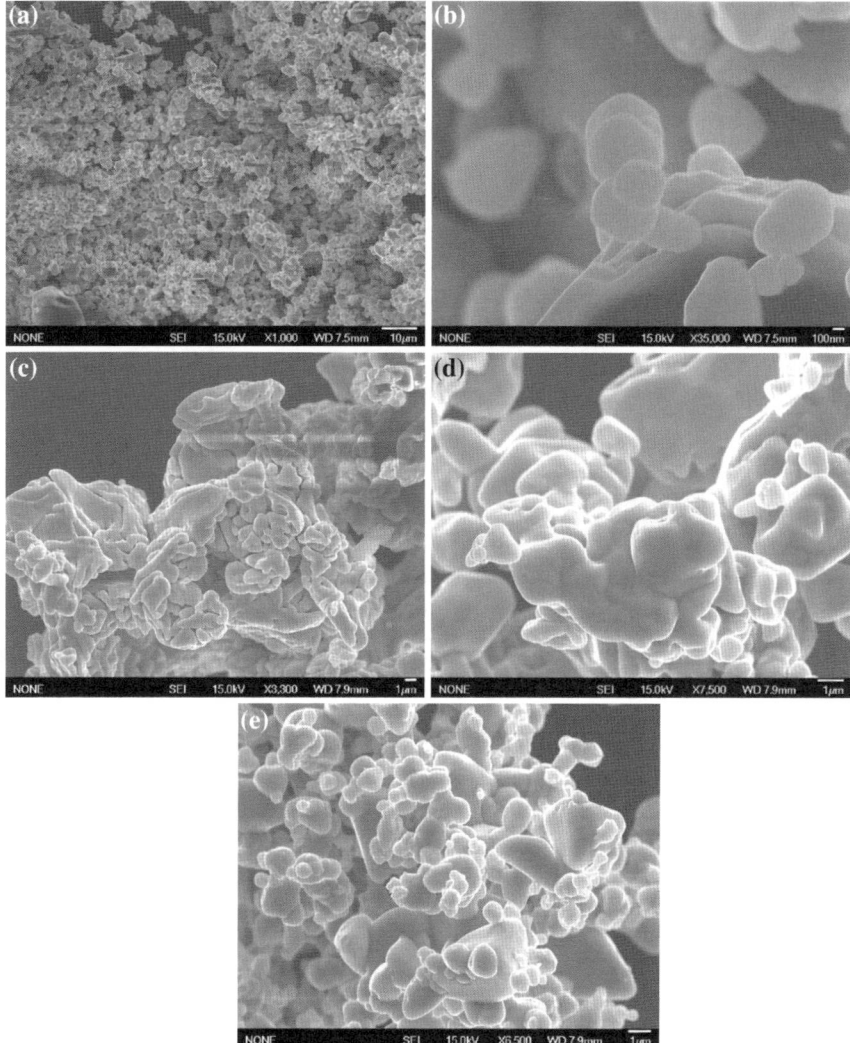

Fig. 3.4 SEM micrographs of samples B1 (**a**, **b**), B2 (**c**, **d**), and B3 (**e**)

The oxygen evolution rate of both B1 and B3 (after 1 h; 1357 and 1159.4 μmol h^{-1} g^{-1}), using Na$_2$HPO$_4$ as a precursors, and with the addition of a small amount of non-aqueous solvent, is analogous to that reported by Yi et al. (1272 μmol h^{-1} g^{-1}). It appears that the precursor must be anhydrous Na$_2$HPO$_4$, and be kept anhydrous with the use of an alcohol to keep the crystal structure phase pure, and OER high. Despite B1 having a slightly smaller surface area in comparison to B3 (1.01 vs. 1.29 m^2 g^{-1}), the OER rate is higher, again suggesting that the activity of Ag$_3$PO$_4$ is not dramatically influenced by surface area.

Fig. 3.5 XRD patterns for samples *B1*, *B2* and *B3*. The standard ISCD/JCPDS pattern for Ag_3PO_4 is shown as *red bars*. The Bragg peaks of the unknown phase are indicated by the symbol '⊗'

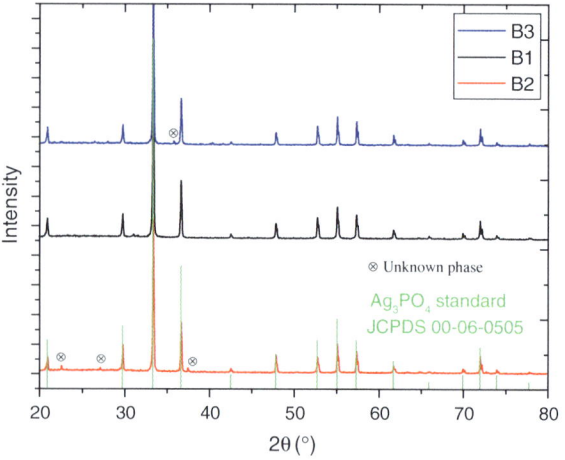

Fig. 3.6 UV-Vis absorbance spectra for samples *B1*, *B2* and *B3*

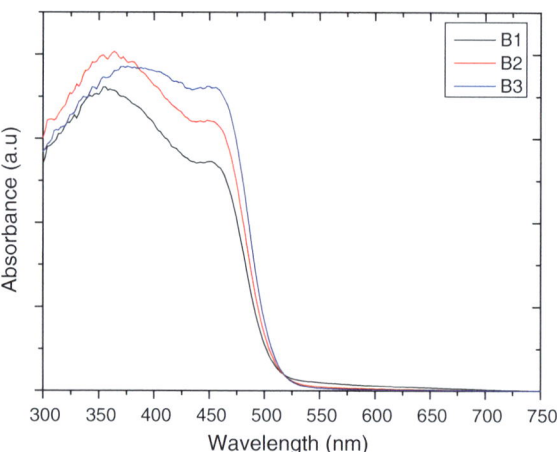

Fig. 3.7 Oxygen evolution under visible light ($l > 420$ nm) from water, in the presence of $AgNO_3$ (0.85 g), at neutral pH, using 0.2 g photocatalyst (*B1*, *B2*, *B3*)

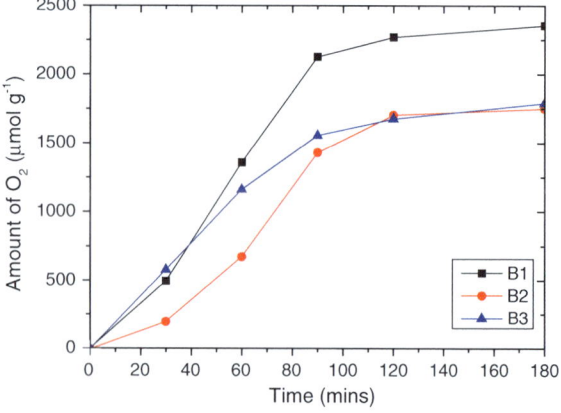

Both low and high resolution SEM micrographs of samples C1 show that the compound has a similar roughly spherical morphology to that of samples synthesised by the previous two methods, 'A' and 'B' (Fig. 3.8). The compound has an almost identical absorbance spectrum (edge at approximately 506 nm) to all previously synthesised samples (Fig. 3.9), and is also pure phase BCC Ag_3PO_4 (Fig. 3.10). However, Fig. 3.11 illustrates that the compound has a higher OER than any other sample, at 1418.75 μmol h^{-1} g^{-1}, exceeding that reported by Yi et al.

Whilst it cannot be proven directly, it would appear to be by using a larger volume of ethanol during synthesis, and thereby having a more dilute solution, is beneficial in synthesising a compound which has little or no unreacted species on its surface. What can be said however, is that the BET SSA does not appear to influence the OER on any Ag_3PO_4 photocatalyst sample. Sample C1 demonstrates the largest OER, yet has the smallest BET SSA (Table 3.2). Although no definitive conclusion can be drawn as to an OER determinate for Ag_3PO_4, it can be said

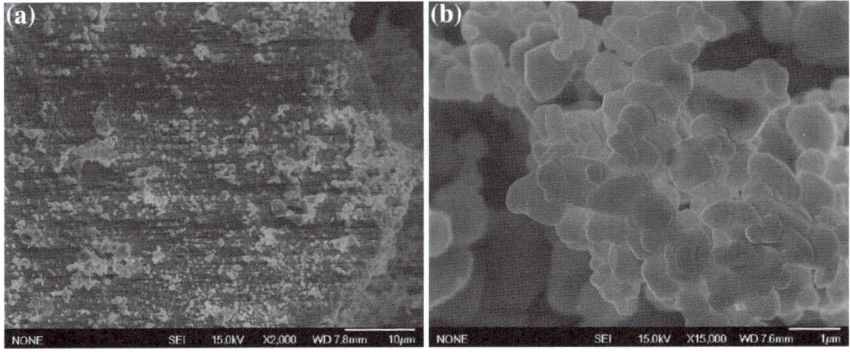

Fig. 3.8 SEM micrographs of sample C1 in low (**a**) and high (**b**) resolution

Fig. 3.9 UV-Vis absorbance spectra of sample *C1*. The grey area visually demonstrates the range of the 420 nm long pass filter. A *black line* indicates the band edge and approximate absorption limit

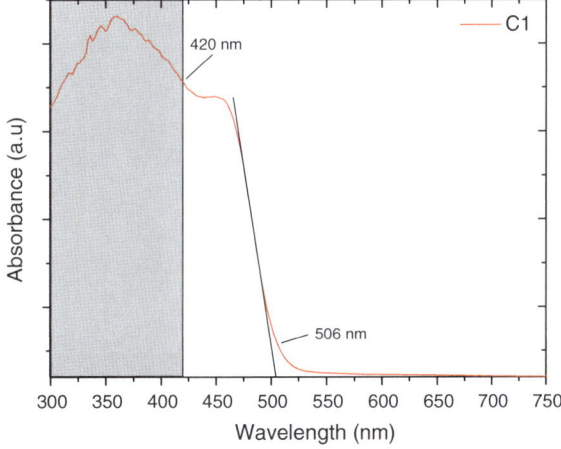

Fig. 3.10 XRD pattern for sample *C1*. The standard ISCD/JCPDS pattern for Ag$_3$PO$_4$ is shown as *red bars*, and the corresponding miller indices of each plane are indicated above each peak

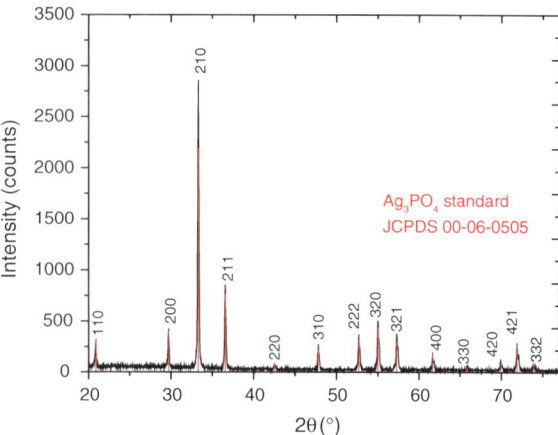

Fig. 3.11 Oxygen evolution under visible light ($l > 420$ nm) from water, in the presence of AgNO$_3$ (0.85 g), at neutral pH, using 0.2 g photocatalyst (*C1*). Indicated are the regions of most importance; the initial period where O$_2$ evolution rate is high (rates are calculated from this point), and where the rate deviates from linearity in the later stages of the experiment

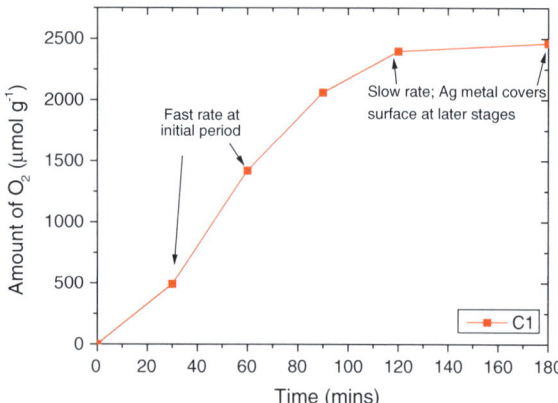

Table 3.2 Properties of different Ag$_3$PO$_4$ crystals

Sample	Facet	Surface energy (J m^{-2})	Hole mass (m_h^*)	Surface area (m^2 g^{-1})	Initial O$_2$ evolution rate (μmol h^{-1} g^{-1})	TOF min^{-1} $\times 10^{-2}$
Tetrahedron	111	1.67	1.53	1.0	6074	2.60
Cube	100	0.67	1.92	1.4	605	0.52
Rhombic dodecahedron	110	0.78	2.08	2.4	426	0.17

that the purity of the crystal phase of the compound is paramount, and also that the precursor and solvent used in synthesis is very important. However, it is now possible to synthesise high purity Ag$_3$PO$_4$ which demonstrates activity for oxygen evolution from water which is comparable with that shown to be possible by Yi et al. [1]. The slight difference between the active sample (C1) synthesised in this study, and that by Yi et al. could be from the different gas chromatographs used, or from the different systems (Yi et al. uses a vacuum system, whilst the system in this report is an Argon atmosphere at 1 bar). Much less likely however, dissolved oxygen could still be contained in the system, which has not been flushed out by the argon purging procedure.

3.3.2 Facet Control of Ag₃PO₄ (Method 'D') [2]

The previous section demonstrated that synthesis parameters of have a significant impact on the OER of Ag_3PO_4 irrespective of surface area. As mentioned previously, specific exposing facets often dominate activity of a catalyst, including TiO_2 for water splitting and Ag_3PO_4 for organic decomposition [12, 17]. Therefore collaborators, Naoto Umezawa and Jinhua Ye, performed surface energy calculations of {100}, {110}, and {111} planes, using a DFT+U approach, and found that out of the three planes, {111} possessed the largest surface energy. Despite of the importance of the exposing facets, no data was available for whether or not the exposing facet influenced the photocatalytic rate of oxygen evolution from water, this line of investigation was pursued.

The surface structures generated by the DFT+U method are based on a slab model including 192 atoms of Ag_3PO_4 with 10 Å thickness of a vacuum region. Geometry relaxations were performed with the mid layers of the slab fixed. The relaxed geometries of the three surfaces are depicted in Fig. 3.12. The surface energy (γ) was calculated from the following formula:

$$\gamma = \frac{(E_{slab} - NE_{bulk})}{2A} \qquad (3.4)$$

where E_{slab} is the total energy of the slab and E_{bulk} is the total energy of the bulk per unit. N and A are the number of units included in the slab and the surface area. The obtained surface energies are 0.67, 0.78, and 1.65 J/m^2 for {100}, {110}, and {111} respectively.

The results clearly show that the {111} facet is over two times more energetic than the second highest facet, {110}. Previous studies by Umezawa et al. about Ag_3PO_4 also show an anisotropic distribution of electron-hole masses (m_h*, Table 3.2) [18]. According to earlier reports, smaller values of m_h* lead to a greater hole mobility; therefore making transport to the surface much more straightforward, and thus decreasing the probability of recombination [19]. Tetrahedral Ag_3PO_4 crystals purely composed of {111} facets were shown to be theoretically possible using Materials Studio (Fig. 3.13).

A kinetically controlled approach in the absence of surfactants operated at ambient temperature was employed (Method 'D'), and crystals with tetrahedral morphology and preferred {111} facets were fabricated in high yield (Fig. 3.15). The exposing {111} facets were controlled by manipulating the concentration of H_3PO_4 in ethanol; whereby excess amounts of phosphoric acid not only provide a precursor, but also control the rate of Ag_3PO_4 nucleation and initial growth (seed

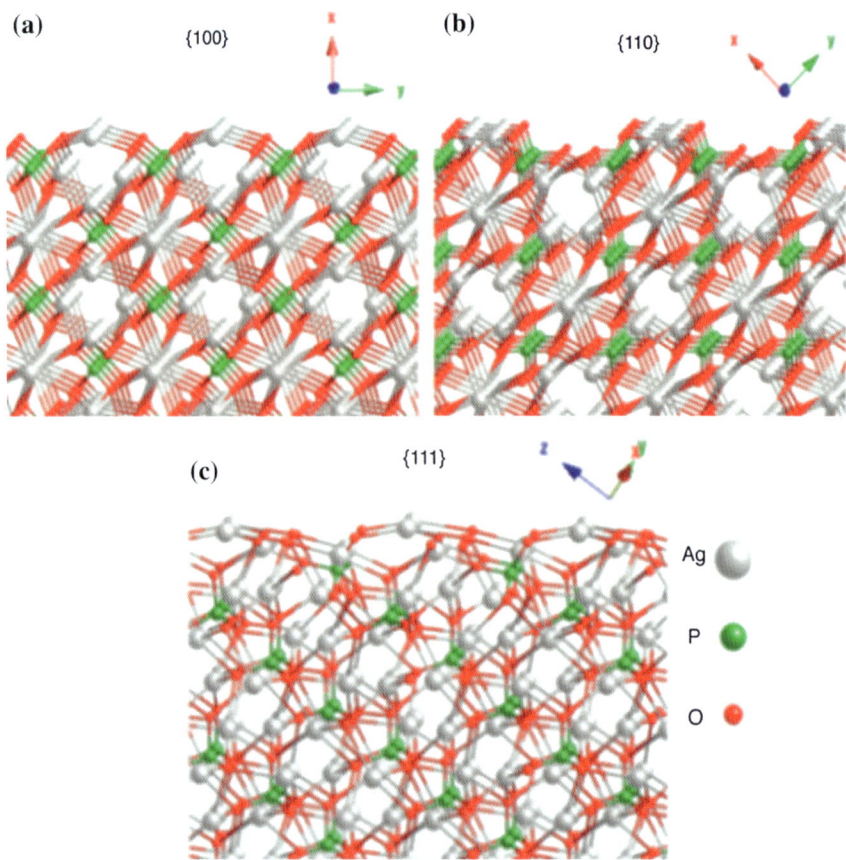

Fig. 3.12 Relaxed geometries for **a** {100}, **b** {110}, and **c** {111} surface of Ag$_3$PO$_4$ (calculations performed by Naoto Umezawa, NIMS) [2]

formation). As the H$_3$PO$_4$ volume is increased from 2 to 15 cm^3, Ag$_3$PO$_4$ tetrapods grow into tetrahedrons through a most probable Oswald ripening effect, as more H$_3$PO$_4$ is available at larger concentrations, and thus smaller particles/precursors will eventually form tetrahedrons, from tetrapods. Directional growth is halted at a specific concentration of H$_3$PO$_4$ in comparison to the silver precursor (AgNO$_3$). No other sub-structures were found either within the powdered samples, or on the clean surfaces of the particles (Fig. 3.14).

Using method 'C', mixed faceted shapes are formed as previously reported [1]. Rhombic dodecahedrons {110} and cubes {100}, which have been reported for organic dye photodecomposition [12], were also fabricated to explore the correlation of facets with photocatalytic activities for oxygen production from water. The different morphologies were also been confirmed by SEM micrographs (Fig. 3.14). In order to further verify the tetrahedral morphology, TEM tilt studies

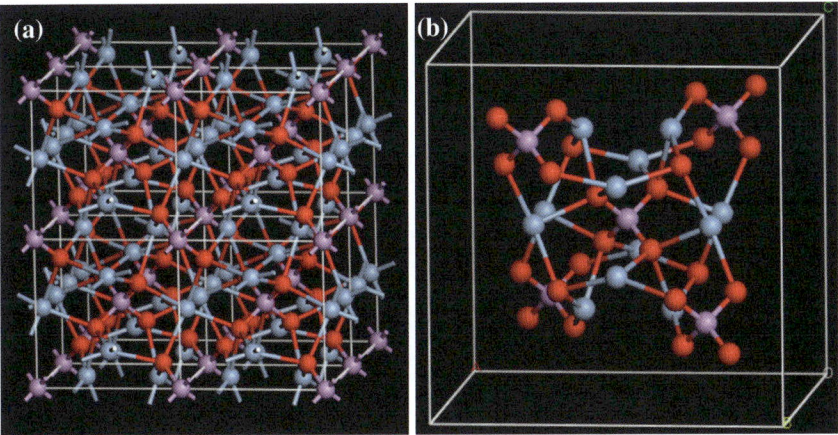

Fig. 3.13 **a** Ag$_3$PO$_4$ 2 × 2 × 2 supercell and **b** Ag$_3$PO$_4$ tetrahedron cell (Figure from Martin et al. [2])

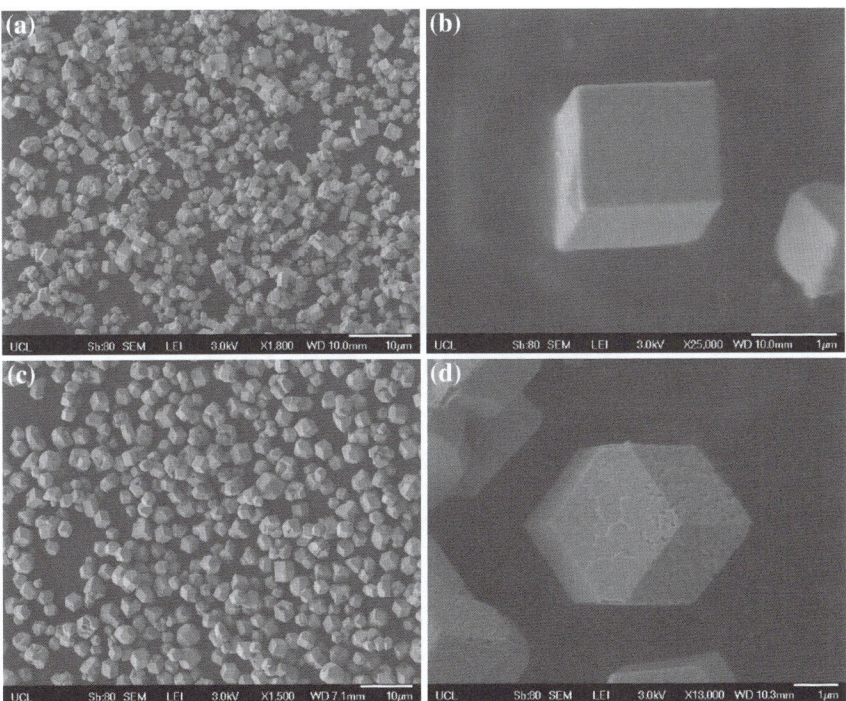

Fig. 3.14 SEM micrographs of silver phosphate cubic crystals (**a, b**), and rhombic dodecahedral crystals (**c, d**)

Fig. 3.15 Low (**a**) and high (**b**, **c**) magnification SEM micrographs of Ag_3PO_4 tetrahedrons (20 cm³ H_3PO_4). Low (**d**) and high (**e**) magnification SEM micrographs of Ag_3PO_4 tetrapods synthesised using 2 cm³ H_3PO_4. **f** SEM micrographs illustrating the formation of Ag_3PO_4 tetrahedrons from tetrapods by varying the concentration of H_3PO_4 (*left* to *right*; 2, 5, 10, 15, 20 cm³ H_3PO_4)

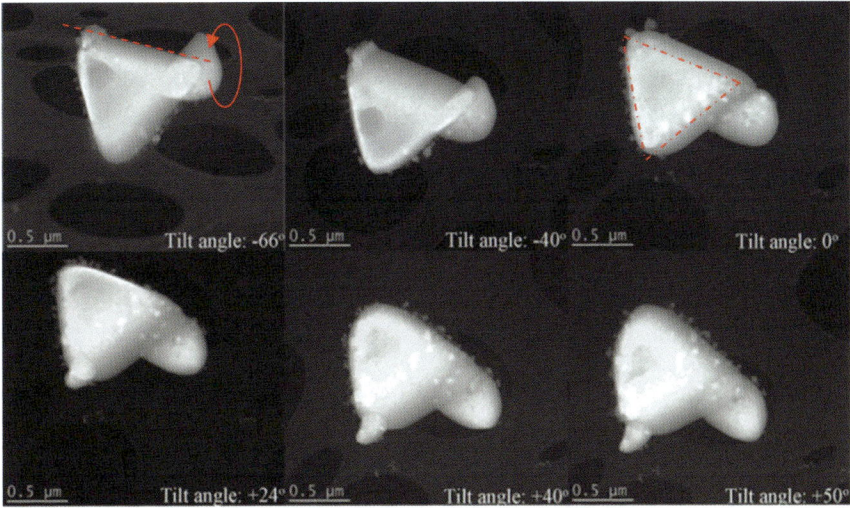

Fig. 3.16   TEM micrograph tilt study, performed by collaborator, Xiaowei Chen. A tetrahedron indicated by *dotted lines* is rotated on axis from −66 to +50°

Fig. 3.17   XRD pattern of faceted Ag_3PO_4 crystals

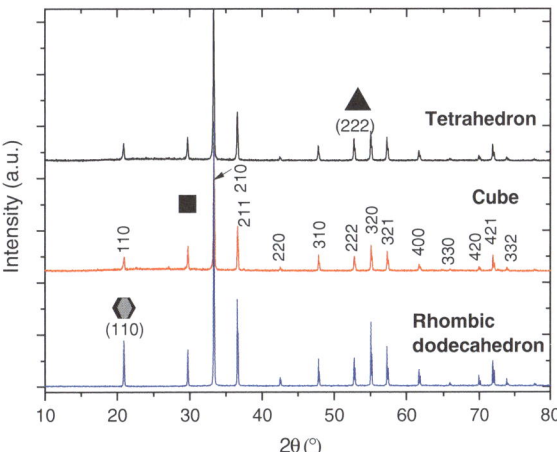

have carefully been undertaken. As seen in Fig. 3.16, the faceted compound is rotated on axis, to reveal 4 identical facets in total from different angles; proving they are tetrahedrons with {111} surfaces.

The X-ray Diffraction patterns of the three samples (Fig. 3.17) indicate a full phase match to that of body-centred cubic crystalline Ag_3PO_4 (6.004 Å, BCC, P-43n, JCPDS no.06-0505), and show no other trace peaks associated with the precursors used in fabrication, indicating that the synthesis methods lead to a completely well crystallised pure phase compound. The diffraction patterns exhibit stark variations in intensity ratios. Tetrahedral Ag_3PO_4 particles have an intensity ratio of 0.98 between (222) and (200) planes, whilst the rhombic dodecahedron and cubic structures have an approximate ratio of 0.68, confirming that tetrahedral particles are comprised of {111} crystalline planes [20]. The intensity ratio between (110) and (200) for rhombic dodecahedrons is 1.23, and 0.57 for the cubic crystals, confirming as previously reported that rhombic dodecahedrons and cubic structures are composed of {110} and {100} facets respectively [21]. It is noted that the (111) Bragg peak is not directly seen due to the reflection conditions of the space group, similarly, (100) cannot be seen either.

Before testing the catalytic activity for direct water photooxidation, the UV-Vis absorption spectra between the three different morphologies were measured (Fig. 3.18). There is a large difference between the samples; rhombic dodecahedrons are able to absorb light up to approximately 530 nm, whilst both tetrahedrons and cubes have an absorption edge around 515 nm. Since the size of the crystals is similar (see Table 3.3 and Fig. 3.19) and there are no signs of foreign elemental doping in the sample, it is thought that the absorption edge shift is likely to be due to the faceted morphology. It is also possible that the absorbance shift is due to a particle quantum size effect (continuous to discrete band shift due to fewer energy levels). However this is unlikely as the particles are over 1 μm diameter.

The faceted samples were tested for the photooxidation of water in the presence of an electron scavenger under ambient conditions. As seen in Fig. 3.20a, oxygen evolves extremely fast, at an initial rate of 6072 $\mu mol\ h^{-1}\ g^{-1}$ for tetrahedral Ag_3PO_4 particles under full arc irradiation, over 10 times more than either rhombic dodecahedral or cubic samples which have been reported for organic photodecomposition but not water splitting [12]. Furthermore, the benchmark

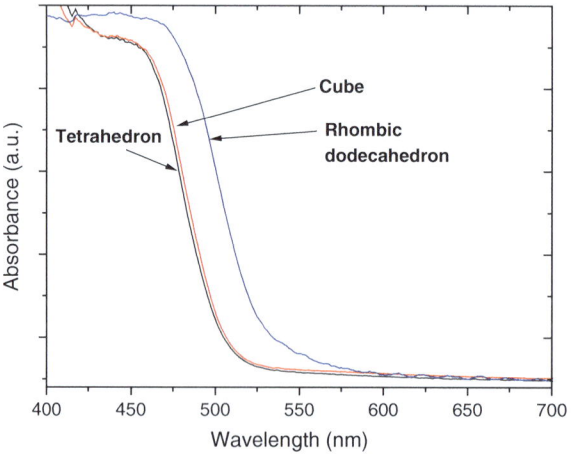

Fig. 3.18 UV-Vis absorbance spectra of faceted Ag_3PO_4 crystals

Table 3.3 Size, range, and standard deviation (S.D.) of various Ag_3PO_4 crystals

Morphology	Mean particle size (μm)	Modal particle size (μm)	Size range (μm)	Standard deviation
Tetrahedron	1.24	1.09	0.70–2.07	0.34
Cube	1.21	1.02	0.51–2.13	0.33
Rhombic dodecahedron	2.91	2.79	1.53–4.40	0.65

Mean, mode, range, and S.D. is calculated for 100 particles using ImageJ

material for visible light water photooxidation, $BiVO_4$, was prepared using a previously described method [22]. Tests under visible light for mixed faceted samples (prepared by method 'C') show a very similar oxygen evolution rate as reported (Fig. 3.20b). {111} faceted Ag_3PO_4 evolves oxygen at a rate of 6.5 times $BiVO_4$ as shown in Fig. 3.20b. The {111} faceted sample shows 242 % more activity in comparison to a mixed faceted sample synthesised using method 'C' and reported by Yi et al. [1] (Fig. 3.20). The activity of all samples was also compared under visible light irradiation (l > 420 nm, Fig. 3.20c). Both rhombic dodecahedrons and cubic crystals are less active than the mixed faceted Ag_3PO_4 for water photooxidation, with tetrahedral Ag_3PO_4 performing best under visible light irradiation.

Using the BET method, tetrahedral Ag_3PO_4 crystals were experimentally verified to have a surface area of 1.4 m^2 g^{-1}, somewhat smaller than cubes (1.8 m^2 g^{-1}) and rhombic dodecahedron (2.4 m^2 g^{-1}), where the raw data is shown in Fig. 3.21. All isotherms can be classified as type II, demonstrating a plateau at $P/P_0 < 0.3$ and no hysteresis; evident of low surface area and very little porosity. The order of activity of oxygen evolution is; tetrahedrons, cubes, rhombic dodecahedrons. Tetrahedral and cubic Ag_3PO_4 have a similar particle size, both in terms of mean and mode, and a similar standard deviation. However, the rhombic dodecahedral crystals possess a particle size more than double that of tetrahedral or cubic crystals, and a large variation in size (2.91 $\pm$ 0.65 mm). Despite the larger particle size of the rhombic dodecahedron sample, it possesses a larger surface area, attributed to an increased porosity (see Fig. 3.14d for evidence of pores), and can absorb more light. Therefore there is no direct correlation between the oxygen evolution rate and either light absorbance, particle size, or surface area.

In order to make a more reliable and reproducible assessment of the activity of {111} faceted crystals, they were also prepared and tested again by me in NIMS, Japan using the same reaction conditions. The difference in activity is approximately 5.1 %, which likely arises from the difference in pressure between the two systems. Therefore not only is the synthesis reproducible, but the performance is very similar in different systems as well.

As shown in Table 3.2, the turnover frequency, which was calculated using Eq. 1.9, divided by time, (over a 3 h experiment) is dramatically different between samples, with the largest gulf between tetrahedral and rhombic dodecahedral samples (15 times higher). It should be noted that the TOF min^{-1} is greatly underestimated, as this calculation assumes all atoms are on the surface and involved in

Fig. 3.19 Particle diameter histograms for Ag₃PO₄ **a** tetrahedrons, **b** cubes, **c** rhombic dodecahedrons

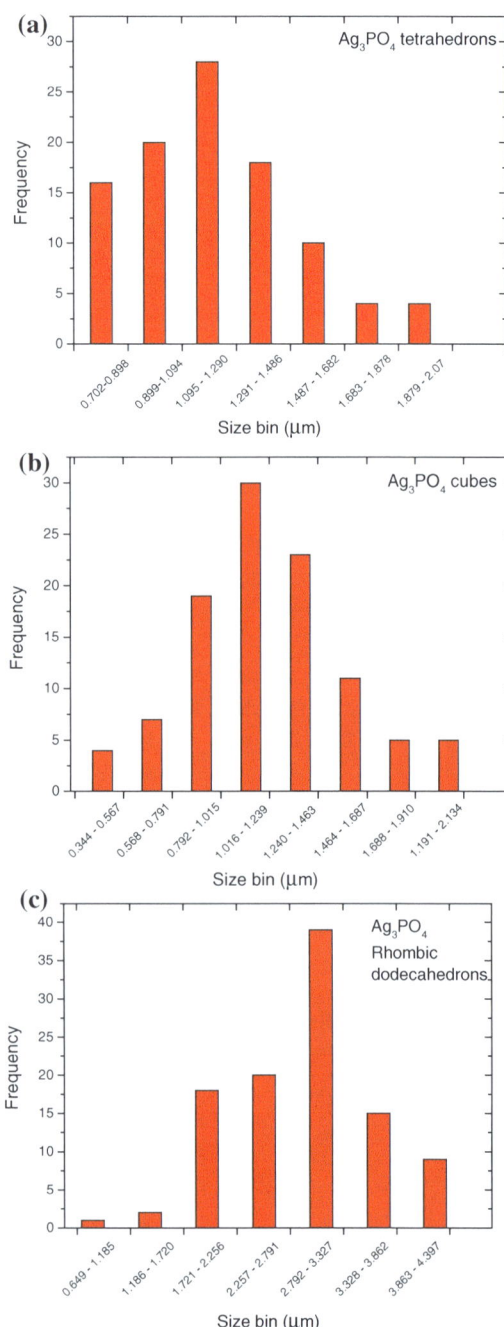

Fig. 3.20 **a** Oxygen yield comparison of Ag_3PO_4 facets using a 300 W Xe light source under full arc irradiation, with $AgNO_3$ acting as an electron scavenger. **b** Oxygen yield comparison between tetrahedral Ag_3PO_4, previously reported random mixed faceted Ag_3PO_4 and $BiVO_4$ under 300 W Xe lamp full arc irradiation, using $AgNO_3$ as an electron scavenger. **c** Oxygen evolution of tetrahedral crystals using a 300 W Xe lamp fitted with 420 nm long pass filter, $AgNO_3$ was used as an electron scavenger (0.85 g). The mixed facet sample was synthesised using method C and Na_2HPO_4. $BiVO_4$ was synthesised according to a previously reported recipe [22]

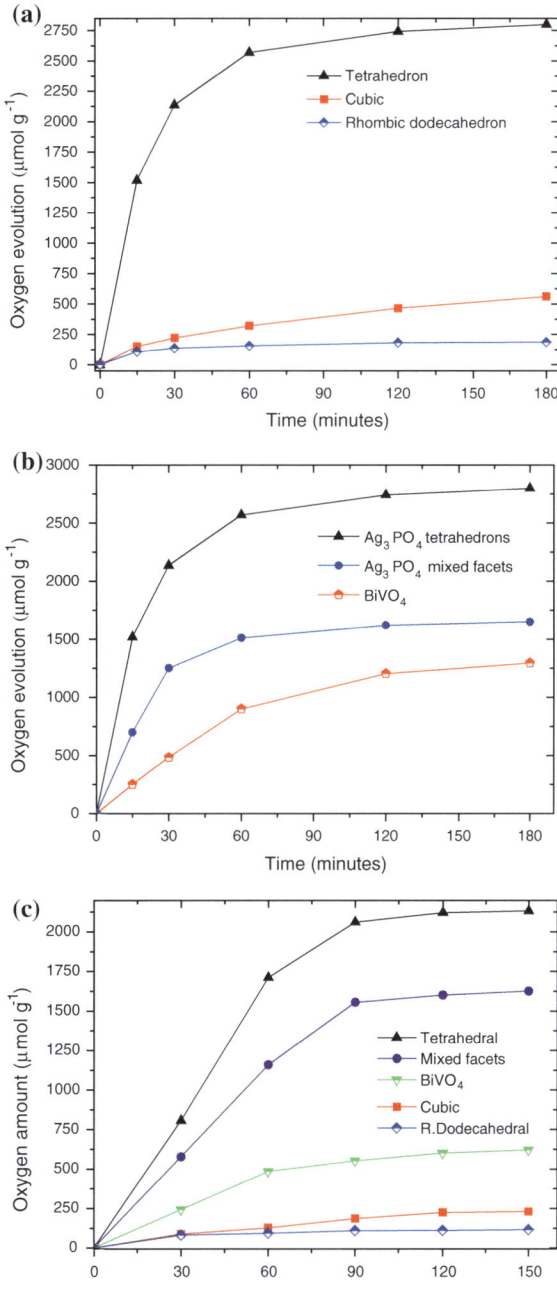

Fig. 3.21 BET adsorption
isotherms for Ag$_3$PO$_4$
tetrahedrons, cubes, and
rhombic dodecahedrons

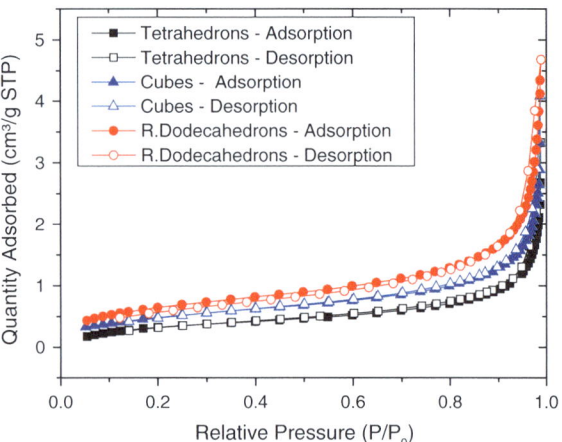

surface reactions, which is not true in this particular case, since the particle size
is about 2 μm, as many atoms do not take part in the photocatalytic surface reaction. The turnover number is ca. 5 for tetrahedral Ag$_3$PO$_4$ particles, meaning that
the reaction is a truly photocatalytic process for the {111} faceted sample (TON
$\geq$ 1 is indicative of a catalytic process). Due to the electron scavenger used in the
experiments, after a period of around 120 min, silver metal (Ag$^+$ + e$^-$ → Ag^0)
has completely covered the surface of the semiconductor, and significantly prevents further oxidation by blocking incident photon flux, as seen previously, and in
other studies [23], which limits the calculated minimum TOF min^{-1} and turnover
number. After the experimental run, the sample was characterised again by XRD.
The diffraction pattern of Ag$_3$PO$_4$ even after an extensive TOF min^{-1} measurement is very similar to the freshly synthesised sample, and as expected, metallic
silver was also detected as a dominant surface species (Fig. 3.22).

Fig. 3.22 XRD pattern
before and after a full arc
photocatalytic test for the
photooxidation of water by
tetrahedral Ag$_3$PO$_4$. AgNO$_3$
was used as a scavenger, and
thus Ag metal deposited on
the surface (JCPDS 00-04-
0783)

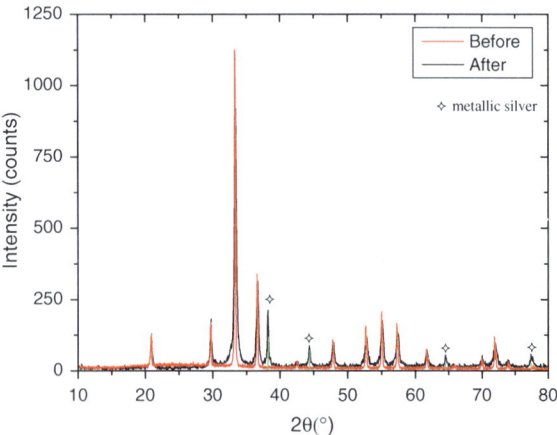

The internal quantum yield for water photooxidation was measured to further demonstrate the efficiency of the tetrahedral Ag_3PO_4 particles. A high quantum yield was observed; 95 % at 420 nm and nearly unity at 400 nm (Fig. 3.23). More importantly, the tetrahedral particles show high quantum yield (>80 %) at a wide wavelength range; from the UV portion of the spectrum throughout the visible until ca. 500 nm, which nearly matches the peak light intensity of solar spectrum (Fig. 3.23b inset). Taking into account the light scattering by the large particles which was assumed zero in our measurement, the actual quantum

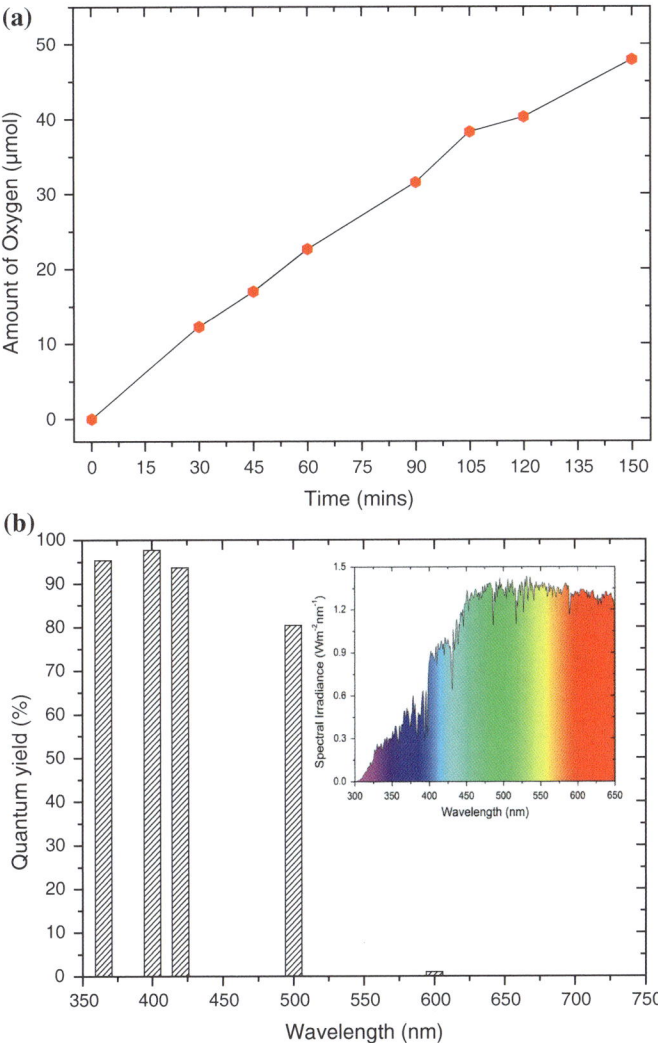

Fig. 3.23 **a** Oxygen evolution using a 400 nm band pass filter, and **b** Internal quantum yield variation using different band pass filters, using $AgNO_3$ as a scavenger, and a 300 W Xe light source

yield could be higher than the reported. This suggests the enhanced photocatalyst is able to harvest light effectively, separate and then utilise charge carriers extremely efficiently.

A suggested mechanism for enhanced photocatalytic activity on the faceted compounds states that depending on the surface structure, electrons or holes will specifically migrate to separate sites [24]. This is due to the differences in the effective masses of holes and electrons, which are directionally sensitive to the crystal structure. The hole mass along the [111] direction is much smaller compared to other directions in Ag_3PO_4 crystals (Table 3.2), indicating that hole mobility and migration towards {111} facets is greatly encouraged, which is a critical issue in an n-type semiconductor such as silver phosphate in which holes are the minor species. The water splitting process is dominated by water oxidation as mentioned in the introduction, and thus a higher mobility of holes favours a faster and more efficient transport of holes to the surface for oxidation reactions. The reason for the high surface energy of the {111} surface is the over-abundance of dangling phosphorus-oxygen bonds, which act as oxidation sites. According to calculations performed by Umezawa, in order to reveal the {111} plane, P-O bonds must be cut, resulting in very short layer by layer distances along [111]. From extensive studies in catalysis [25], it is widely accepted that a high surface energy is very beneficial to surface reaction rates. The high surface energy dramatically enhances the catalytic ability at the surface of {111} facets for example, due to more step sites and kinks. The combination of both a high hole mobility and high surface energy results in a unique synergistic effect, culminating in a significant increase in water photooxidation. Both Ag_3PO_4 rhombic dodecahedrons and cubes possess slightly higher surface areas, but they have larger hole masses and much smaller surface energies, which leads to a lower activity. It has been reported that both rhombic dodecahedrons and Ag_3PO_4 cubes show better activity than the mixed faceted Ag_3PO_4 for organic dye photobleaching [12]. This chemical process is mechanistically quite different from water splitting. Thermodynamically, there is a difference in required chemical potentials between water splitting and dye bleaching. More importantly, kinetically as a dominating step in water splitting, water photooxidation requires 4 holes, and occurs on relatively long timescale of ca. 0.5 s [4, 6]. However the dominating step in organic photooxidation is oxygen reduction reaction, which is relatively fast on microsecond timescale [4–6, 26]. Historically for example, n-doped TiO_2 is able to bleach organic contaminants faster than pure TiO_2, but unable to efficiently evolve molecular oxygen under visible irradiation, even in the presence of scavengers [5]. Analogously, both rhombic dodecahedrons and Ag_3PO_4 cubes perform much worse for water photooxidation, which is probably due to their high number of unfavourable surface reaction sites. The mixed faceted standard Ag_3PO_4 sample presents an intermediate activity, which is expected considering there are a mixture of all exposing facets.

3.4 Conclusions

Initially, different sodium based phosphate precursors were used to synthesise three series of Ag_3PO_4 crystals all with irregular morphology. Furthermore, synthesis was also carried out in the absence, and presence of water and ethanol as ion exchange solvents. These initial samples (series A, B and C) demonstrated varying abilities to photooxidize water, yet there was no correlation between particle size or BET surface area. Indirectly therefore, it was deduced that using large volumes of ethanol yielded a precipitate with a cleaner surface, containing little or no surface species such as sodium or atomic hydrogen. Anhydrous precursors also demonstrated better OERs in comparison to compounds whose precursors contained water. Despite this, it was possible to synthesise a sample (C1) which demonstrates an activity comparable to the study by Yi et al. (in fact, it was better by ca. 20 %).

In light of collaborators' DFT calculations, it was then decided that controlling the exposing facet of Ag_3PO_4 crystals could potentially lead to an understanding over what determines the governing factor of oxygen evolution from water. It was already apparent from series A, B and C that the BET surface area did not influence the OER.

Using method C as a starting point, tetrahedral Ag_3PO_4 crystals composed of {111} terminated surfaces were synthesised by using H_3PO_4 to control the reaction kinetics of crystal growth. The experimental results show that the synthesised tetrahedrons not only exhibit nearly unity internal quantum yield at 400 nm but also boast a high quantum yield (>80 %) over a very wide wavelength range, from the UV to 500 nm. The activity for water photooxidation is nearly 12 times greater than {100} and {110} faceted crystals which were synthesised using a modified method based on the work of Bi et al. Tetrahedral Ag_3PO_4 crystals also show 240 % and 650 % enhancement compared with benchmark photocatalysts, Ag_3PO_4 (mixed facets, sample C1) and $BiVO_4$, respectively. In addition, the {111} terminated crystals show an underestimated TOF min^{-1} of 2.6×10^{-2} and turnover number of ca. 5, even though the particle size is very large (>2 mm), and surface area is very small. The excellent performance of {111} faceted Ag_3PO_4 tetrahedral crystals is attributed a cooperative effect between high mobility of holes and high surface energy.

The hole mass along the [111] direction is much smaller compared to both {110} and {100} terminated Ag_3PO_4 crystals, demonstrating that hole migration without recombination towards the {111} facet is statistically more likely to occur. Since water oxidation is the rate determining step in water splitting, and occurs on a slow timescale (ms), the likelihood of electron hole recombination is much larger if the hole mass is larger/hole mobility is smaller. Thus a higher mobility of holes favours a faster and more efficient transport of holes to the surface and water for oxidation reactions. The high surface energy then dramatically enhances the catalytic ability at the surface of {111} facets, due to more step sites and kinks as shown in the tetrahedral morphology. Whilst not being able to exactly identify the number of kinks or steps, on closer inspection of Fig. 3.12, it is apparent that there is an absence of phosphorus bonds at the {111} surface, a result of cleaving P-O bonds when a {111} facet is formed. It is also known that the active site for water photooxidation is the oxygen atom—hence there are more oxygen atoms per unit surface area, due to the absence of phosphorus. Therefore the presence of phosphorus at the surface might be detrimental to the photooxidation of water.

In general, the findings provide an effective way to improve the efficiency of a photocatalyst; by controlling the exposing facet. The synthesis preparation method also provides useful information for a reproducible, surfactant free and facile way to control surface facet and bulk properties. With relation to the aim of the thesis, the chapter demonstrates the facet effect for water photooxidation, which dramatically increases OER and internal quantum yield in comparison to non-faceted crystalline photocatalysts.

An opposing argument for the importance of developing a water photooxidation, which is viewed as the key to water splitting due to the temporal difficulties, is the development of a photocatalyst which is able to produce hydrogen from water under visible light. A key problem seen in many photocatalysts is that the positioning of the stable O2p orbital is so deep (located around $+3$ eV), that visible light absorption is limited if the compound also possesses a conduction band negative enough to reduce protons to H_2. It is possible to dope the band structure to add sub band gap states and create visible light excitation centres (e.g. Rh or Cr doped $SrTiO_3$), however this approach can never achieve quantum yields high enough required for hydrogen synthesis due to the lack of electronic states present at a acceptor/donor level. Therefore, considering the surveyed literature, the next chapter will thus focus on the feasibility and advancement of a newly discovered, undoped, stable photocatalyst for hydrogen production from water under visible light, graphitic carbon nitride.

References

1. Yi, Z., et al. (2010). An orthophosphate semiconductor with photooxidation properties under visible-light irradiation. *Nature Materials, 9*, 559–564.
2. Martin, D. J., Umezawa, N., Chen, X., Ye, J., & Tang, J. (2013). Facet engineered Ag_3PO_4 for efficient water photooxidation. *Energy & Environmental Science, 6*, 3380–3386.
3. Reece, S. Y., et al. (2011). Wireless solar water splitting using silicon-based semiconductors and earth-abundant catalysts. *Science, 334*, 645–648.
4. Pendlebury, S. R., et al. (2011). Dynamics of photogenerated holes in nanocrystalline α-Fe_2O_3 electrodes for water oxidation probed by transient absorption spectroscopy. *Chemical Communications 47*, 716–718.
5. Tang, J., Cowan, A. J., Durrant, J. R., & Klug, D. R. (2011). Mechanism of O_2 production from water splitting: nature of charge carriers in nitrogen doped nanocrystalline TiO_2 films and factors limiting O_2 production. *The Journal of Physical Chemistry C, 115*, 3143–3150.
6. Tang, J., Durrant, J. R., & Klug, D. R. (2008). Mechanism of photocatalytic water splitting in TiO_2. Reaction of water with photoholes, importance of charge carrier dynamics, and evidence for four-hole chemistry. *Journal of the American Chemical Society, 130*, 13885–13891.

7. Buckley, J. J., Lee, A. F., Olivi, L., & Wilson, K. (2010). Hydroxyapatite supported antibacterial Ag₃PO₄ nanoparticles. *Journal of Materials Chemistry, 20*, 8056–8063.
8. Hu, H., et al. (2013). Facile synthesis of tetrahedral Ag₃PO₄ submicro-crystals with enhanced photocatalytic properties. *Journal of Materials Chemistry A, 1*, 2387–2390.
9. Yang, H. G., et al. (2008). Anatase TiO₂ single crystals with a large percentage of reactive facets. *Nature, 453*, 638–641.
10. Xia, Y. N., Xiong, Y. J., Lim, B., & Skrabalak, S. E. (2009). Shape-controlled synthesis of metal nanocrystals: simple chemistry meets complex physics? *Angewandte Chemie-International Edition, 48*, 60–103.
11. Wang, H., et al. (2012). A facile way to rejuvenate Ag₃PO₄ as a recyclable highly efficient photocatalyst. *Chemistry – A European Journal, 18*, 5524–5529.
12. Bi, Y., Ouyang, S., Umezawa, N., Cao, J., & Ye, J. (2011). Facet effect of single-crystalline Ag₃PO₄ sub-microcrystals on photocatalytic properties. *Journal of the American Chemical Society, 133*, 6490–6492.
13. Mills, A., & Valenzuela, M. A. (2004). The photo-oxidation of water by sodium persulfate, and other electron acceptors, sensitised by TiO₂. *Journal of Photochemistry and Photobiology A: Chemistry, 165*, 25–34.
14. Murphy, A. B., et al. (2006). Efficiency of solar water splitting using semiconductor electrodes. *International Journal of Hydrogen Energy, 31*, 1999–2017.
15. Serpone, N., Terzian, R., Lawless, D., Kennepohl, P., & Sauvé, G. (1993). On the usage of turnover numbers and quantum yields in heterogeneous photocatalysis. *Journal of Photochemistry and Photobiology A: Chemistry, 73*, 11–16.
16. Chen, Z., et al. (2010). Accelerating materials development for photoelectrochemical hydrogen production: Standards for methods, definitions, and reporting protocols. *Journal of Materials Research, 25*, 3–16.
17. Wang, J., et al. (2012). Facile synthesis of novel Ag3PO4 tetrapods and the 110 facets-dominated photocatalytic activity. *CrystEngComm, 15*, 39–42.
18. Umezawa, N., Shuxin, O., & Ye, J. (2011). Theoretical study of high photocatalytic performance of Ag₃PO₄. *Physical Review B, 83*, 035202.
19. Yan, S., et al. (2013). An ion exchange phase transformation to ZnGa₂O₄ nanocube towards efficient solar fuel synthesis. *Advanced Functional Materials, 23*, 758–763.
20. Wang, Z. L. (2000). Transmission electron microscopy of shape-controlled nanocrystals and their assemblies. *J Phys Chem B, 104*, 1153–1175.
21. Yang, J., Qi, L., Lu, C., Ma, J., & Cheng, H. (2005). Morphosynthesis of rhombododecahedral silver cages by self-assembly coupled with precursor crystal templating. *Angewandte Chemie International Edition, 44*, 598–603.
22. Kudo, A., Omori, K., & Kato, H. (1999). A novel aqueous process for preparation of crystal form-controlled and highly crystalline BiVO₄ powder from layered vanadates at room temperature and its photocatalytic and photophysical properties. *Journal of the American Chemical Society, 121*, 11459–11467.
23. Mills, A., & Valenzuela, M. A. (2004). Photo-oxidation of water sensitized by TiO₂ and WO₃ in presence of different electron acceptors. *Revista Mexicana de Fisica, 50*, 287–296.
24. Liu, G., Yu, J. C., Lu, G. Q., & Cheng, H.-M. (2011). Crystal facet engineering of semiconductor photocatalysts: motivations, advances and unique properties. *Chemical Communications, 47*, 6763–6783.
25. Zhou, Z.-Y., Tian, N., Li, J.-T., Broadwell, I., & Sun, S.-G. (2011). Nanomaterials of high surface energy with exceptional properties in catalysis and energy storage. *Chemical Society Reviews, 40*, 4167–4185.
26. Chen, C., Ma, W., & Zhao, J. (2010). Semiconductor-mediated photodegradation of pollutants under visible-light irradiation. *Chemical Society Reviews, 39*, 4206–4219.

Chapter 4
Hydrogen Evolving Photocatalyst Development

This chapter explores the feasibility of g-C_3N_4 as a photocatalyst for hydrogen production from water. Graphitic carbon nitride was synthesised via thermal decomposition, using 4 different precursors, at various calcination temperatures, calcination ramp rates, cleaning techniques, and cocatalyst element/weighting. A novel strategy was found to make highly efficient g-C_3N_4 for H_2 production from urea, resulting in an internal quantum yield for hydrogen production of 26.5 % at 400 nm, over an order of magnitude larger than previously reported values [1]. In order to determine the reason for the high reactivity, in comparison to previously documented examples, a series of characterisation methods were employed. TGA-DSC-MS was used to monitor the formation mechanism of g-C_3N_4 derived from both urea and thiourea—which was found to be drastically different during the primordial stages of synthesis. In the case of urea derived g-C_3N_4, it is shown that the high surface area and porosity is created by the continuous loss of CO_2, a result not seen before. Formaldehyde (CH_2O) is also continually produced throughout the decomposition of urea to g-C_3N_4, a process which is proposed to reduce the percentage of hydrogen/protons in carbon nitride.

Using the N1 s core level data from XPS, it is shown that from sample to sample, there is a correlation between surface hydrogen/proton content and hydrogen production rate. A larger surface hydrogen/proton content leads to a lower hydrogen production rate. This trend is further confirmed by bulk hydrogen analysis carried out by elemental analysis. It has been shown in previous studies that a less polymeric carbon nitride intrinsically has more hydrogen/protons due to breakages, but it was unclear as to why this leads to a less effective photocatalyst. XRD analysis confirms that a greater level of condensation, and thus a less polymerised g-C_3N_4 leads to a less active photocatalyst.

In collaboration with Dr. Stephen Shevlin and Prof. Z. Xiao Guo, using DFT calculations, we demonstrate that increasing the hydrogen/proton content causes a positive shift in the conduction band edge (with respect to NHE), therefore reducing the driving force for reduction reactions. As shown by TDDFT excess protonation localizes photoelectrons at non-active redox sites, which is further detrimental to the photocatalytic ability.

© Springer International Publishing Switzerland 2015
D.J. Martin, *Investigation into High Efficiency Visible Light Photocatalysts for Water Reduction and Oxidation*, Springer Theses,
DOI 10.1007/978-3-319-18488-3_4

4.1 Introduction

Very recently, a stable, organic photocatalyst, graphitic carbon nitride (g-C_3N_4), was found to show sufficient redox power to dissociate water under visible light in a suspension [2]. However the latest documented quantum yields for H_2 production from water using g-C_3N_4 do not exceed 4 % (excluding dye-sensitized systems) [2–11], which is still unsatisfactory for industrial applications [12]. Graphitic carbon nitride is composed of extremely abundant elements and is non-toxic with proven stability, both thermally and in solutions of pH 1–14. Therefore if an effective and facile strategy is devised to improve its energy conversion efficiency, it can potentially meet all the aforementioned three requirements for a practical photocatalyst.

The structure of carbon nitride has been debated since initial reports in 1834; with melon, melem and melam structures all being proposed [13–15]. It is now accepted that the graphitic form of carbon nitride, which is photocatalytically active, is initially composed of stable C-N heptazine (tri-s-triazine) sheets, hydrogen bonded in a zig-zag formation [16]. Graphitic carbon nitride has a rather unique semiconductor structure; the valence band edge is dominated by nitrogen p_z orbitals, and the conduction band edge is dominated by carbon p_z orbitals, with a significant contribution from the p_z orbitals of the edge nitrogen atoms (twofold-bound). These are all contained within one heptazine unit [2]. The heptazine unit connecting planar nitrogen however does not contribute to the valence band edge. Upon further condensation and removal of NH_3 groups, heptazine units continue to polymerise to form what is now conventionally known as graphitic carbon nitride ($C_xN_yH_z$) [17, 18]. However, as shown both experimentally and theoretically, there are different degrees of polymeric condensation; which can greatly influence the activity [4, 17–20]. From previous findings, it seems that it is not the most condensed graphitic carbon nitride which demonstrates the best activity, but the most polymeric; defects are seemingly detrimental to electron and hole localisation at termination sites [2]. Residual hydrogen, which is present in nearly all carbon nitride structures (as a result of incomplete perfect condensation), appears when this connecting amino sp^2 bond is broken, and in principal the amount rises as the polymer begins to buckle. A higher temperature results in in increased condensation; eventually buckling the graphitic structure, and decreasing activity [4]. Up until now however, there has been little direct evidence detailing the activity of g-C_3N_4 for hydrogen production to the structure and in particular surface features of the compound.

4.2 Methodology

4.2.1 Photocatalytic Analysis

Reduction reactions (protons to H_2) were used to analyse the photoactivity of various types of g-C_3N_4. All reactions were carried out in a custom Pyrex batch reactor cell (3.6 cm diameter of reactor window), which was thoroughly purged

Table 4.1   Table showing the actual wt% of some metal precursors

Metal	Precursor	Value of 'x'—number of molecules of H_2O	Actual wt% of metal in precursor (according to certificate of analysis)
Platinum, Pt	$H_2PtCl_6 \cdot (H_2O)_6$	6	37.90
Gold, Au	$HAuCl_4 \cdot xH_2O$	3–4	49.96
Ruthenium, Ru (III)	$RuCl_3 \cdot xH_2O$	≤ 1	39.00

with argon prior to radiation. Gas concentration analysis was performed using a GC (Varian 430-GC, TCD, argon carrier gas 99.999 %).

For a typical reduction reaction, the photocatalyst (0.02 g) was suspended and subsequently sonicated in a deionised water/hole scavenger mixture (230 cm^3 total volume; 210 cm^3 DI water, 20 cm^3 hole scavenger). A co-catalyst was then deposited onto g-C_3N_4 using an in situ photodeposition method [21]. Stock solutions of deionised water and precursors (Pt: $H_2PtCl_6 \cdot (H_2O)_6$, Au: $HAuCl_4 \cdot xH_2O$, Ru : $RuCl_3 \cdot xH_2O$) were made beforehand, and a set volume added according to the required weight of metal (various wt%, Table 4.1). The reactor was sealed, purged with Ar gas for 1 h, and then irradiated for 1 h under full arc irradiation using a 300 W Xe lamp (TrusTech PLS-SXE 300/300UV). During a one hour period, periodic measurements were taken to determine if hydrogen was being produced at a stable rate, and thus, if photodeposition had occurred correctly. The reactor was then purged a second time, prior to irradiation, limited by a 395 nm long pass filter (Comar Optics).

Internal quantum yield (IQY, %) was measured by inserting an appropriate band pass filter (365, 400, 420, 500 nm, $\lambda \pm 10$ nm at 10 % of peak height, Comar Optics) in front of a 300 W Xe light source (TrusTech PLS-SXE 300/300UV) to obtain the correct wavelength. The sample preparation procedure is analogous to a hydrogen evolution measurement and the only difference is 0.1 g photocatalyst is used to ensure complete light absorbance by the photocatalysts. The light intensity was measured at 5 different points to obtain an average intensity, using a photodetector coupled with an optical power meter (Newport Spectra). The intensity of the incoming light was also lowered using a neutral density filter (50 %, Comar Optics), with an average final intensity of 500 ± 25 μW cm^{-2} incident upon the photocatalyst. The efficiency was then calculated using the following formula (ignoring reflected/scattered photons); IQY (%) = (2 × amount of H_2 molecules evolved/number of photons absorbed) × 100. The turnover number (TON) in terms of cocatalyst was calculated after 6 h by using the following formula; TON = moles of H_2 molecules evolved/moles of active sites (platinum on the photocatalyst), for example:

$$TON = \frac{moles\ of\ H_2\ evolved}{moles\ of\ platinum\ on\ photocatalyst}$$

$$= \frac{1.972\ mmol}{0.0006\ g\ (Pt) \times (\frac{1}{195.084\ g\ mol^{-1}})}$$

$$= \frac{1.972 \text{ mmol}}{0.003 \text{ mmol}}$$

$$TON \sim 650$$

4.2.2 Synthesis Techniques

Graphitic carbon nitride was synthesised using different precursors; urea, dicyan-diamide (DCDA), thiourea and cyanamide (Sigma-Aldrich 99.999 %) [19, 20]. In a typical run, the precursor was put in a lidded high quality alumina crucible, then placed inside a muffle furnace and heated at different ramp rate (ranging from 5 to 25 °C/min), and finally held at a designated temperature (ranging from 550 to 650 °C) for several hours. The resultant powders were then washed with water, HCl, NaOH and once again with water to remove all unreacted and potentially detrimental surface species. The products were denoted as g-C_3N_4 (precursor). The procedure was then carefully optimised to give the highest activity for H_2 production from water. Carbon nitride synthesised from melamine and cyanuric acid (MCA) in dimethyl sulfoxide (DMSO), was provided by Kaipei Qiu at UCL Chemistry [22]. Briefly, melamine (3.96 mmol) was dissolved in 20 cm^3 DMSO. Equimolar cyanuric acid was dissolved separately in 10 cm^3 DMSO. The two solutions were kept separate at 30 °C and then mixed for 10 min. MCA-DMSO crystals were then made after storing on a petri dish for 2 days in fume hood. Afterwards, the crystals were heated to 600 °C in nitrogen with a ramp rate of 2.3 °C/min, and after cooling, g-C_3N_4-MCA-DMSO-30-600 was formed.

4.3 Results and Discussion

Graphitic carbon nitride was synthesised from urea, DCDA, thiourea under identical conditions in order to determine to what effect the precursor has on the final product. XRD was first used to investigate and verify the crystal structure of the samples. Figure 4.1a, shows there is a very small difference of in terms of peak position (2θ) in the XRD patterns between samples. The patterns all exhibit peaks at 13.0 (d = 0.681 nm) and 27.4 (d = 0.326 nm), corresponding to the approximate dimension of the tri-s-triazine unit, and the distance between graphitic layers respectively. However on closer inspection, it is evident that the intensity/height of the 27.4 peak varies between sample, following the order urea < thiourea < DCDA (Fig. 4.1b). This implies that the most crystallised/condensed sample is DCDA, and urea the least, as a more intense peak implies that there is more regular repetitions between graphitic layers.

The ATR-FTIR spectra shown in Fig. 4.2 illustrates the typical bands found in graphitic carbon nitrides, with varying intensity, due to the difference in

Fig. 4.1 XRD patterns of
g-C$_3$N$_4$ synthesised using
different precursors using
5 °C ramp rate, 600 °C,
4 h hold time

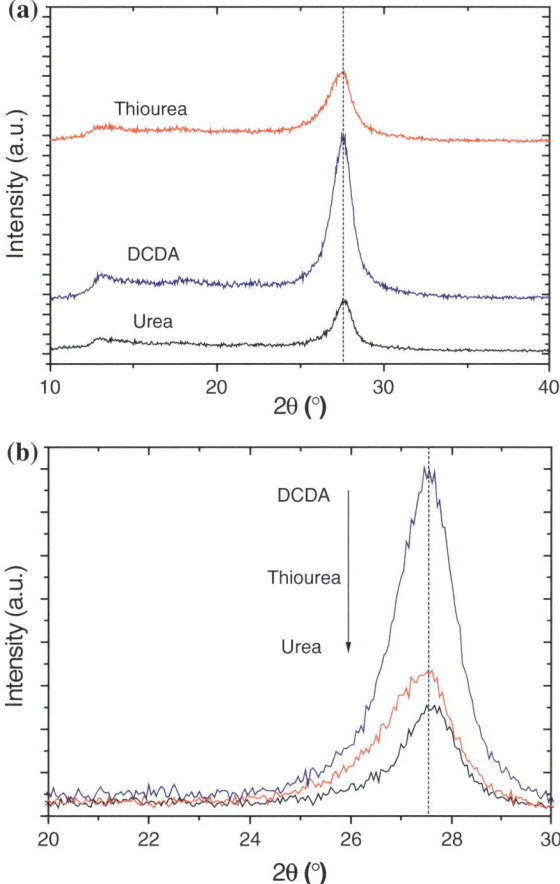

Fig. 4.1 XRD patterns of g-C$_3$N$_4$ synthesised using different precursors using 5 °C ramp rate, 600 °C, 4 h hold time

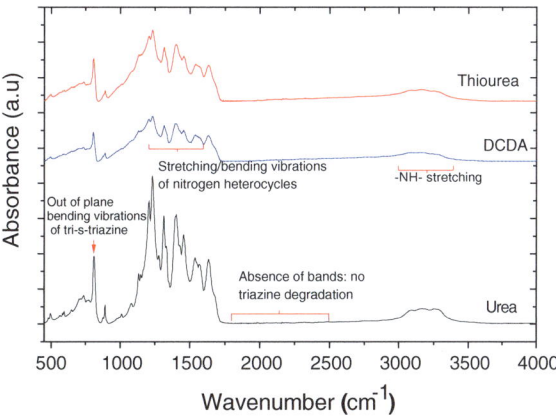

Fig. 4.2 ATR-FTIR spectra of powdered graphitic carbon nitride; annotated are characteristic bands for g-C$_3$N$_4$ [23]

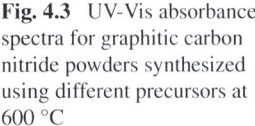

Fig. 4.3 UV-Vis absorbance spectra for graphitic carbon nitride powders synthesized using different precursors at 600 °C

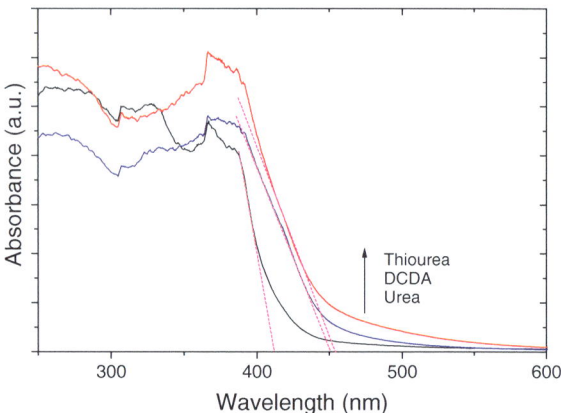

compressibility between samples—physically the urea sample seems less dense and more compressible in comparison to DCDA and thiourea samples (this is due to a larger surface area, discussed later). As annotated, the peaks between 1000 and 1750 cm^{-1} are representative of the stretching and bending modes of nitrogen containing heterocycles, and the broad feature at 3250 cm^{-1} is assigned to stretching modes of -NH- groups. Despite peaks being more distinct in the urea sample, the intensity ratio between -NH- and C-N heterocycles descends in the order DCDA Thiourea urea; indicating that there is possibly less -NH- in g-C$_3$N$_4$ prepared using urea.

Figure 4.3 shows the UV-Vis spectra of the different g-C$_3$N$_4$ samples. All three samples exhibit the characteristic absorption of g-C$_3$N$_4$ at around 380 nm which is assigned to π–π* transitions normally seen in ring systems/heterocyclic aromatics [4]. The shape of the absorption profile is analogous to a semiconductor band gap, with varying approximate band gaps. The sample synthesised from thiourea shows the largest absorption, with an edge around 455 nm. The DCDA derived sample absorbs up to ca. 450 nm, whilst the sample fabricated using urea exhibited a visible blue shift, absorbing light up to approximately 420 nm. All the samples have a small absorption tail, probably due to some sub band gap states from classically forbidden n-π* transitions [4, 24].

Raman spectroscopy allows the fine structure of the three samples to be examined in detail, as shown in Fig. 4.4. From 1200 to 1700 cm^{-1} a series of peaks dominate, attributed to C-N stretching vibrations, specifically G and D band profiles of structurally disordered graphitic carbon and other carbon/nitrogen layered compounds [25, 26]. The peak at 980 cm^{-1} can be assigned to the symmetric N-breathing mode of heptazine, whilst the peak at 690 cm^{-1} corresponds to the in-plane bending vibrations of the heptazine C-N-C linkages [27]. All peaks therefore confirm the local structure of the samples synthesised from various precursors eventually yields the same product; g-C$_3$N$_4$.

Transmission electron microscopy (TEM) images in Fig. 4.5a–c show the highly amorphous nature of the compounds, and HRTEM micrograph (Fig. 4.5d)

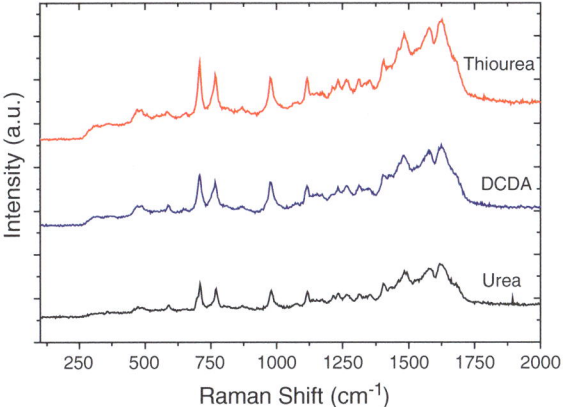

Fig. 4.4   Raman spectra of different carbon nitride samples

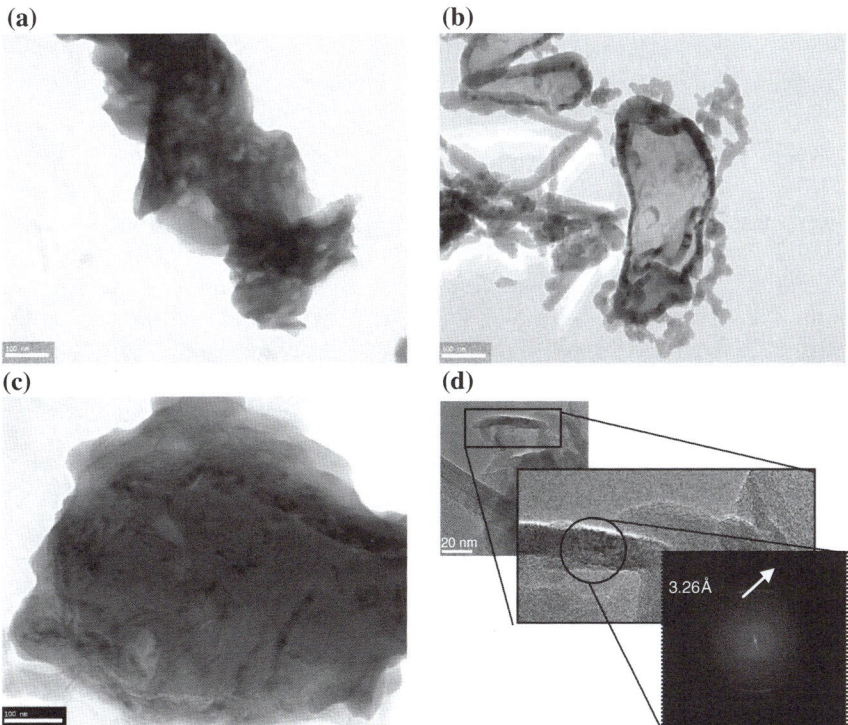

Fig. 4.5   TEM micrographs of g-C$_3$N$_4$ synthesised from: **a** Urea, **b** thiourea, and **c** DCDA. The micrograph scheme at **d** shows a continual magnification of the g-C$_3$N$_4$ (urea) structure, and the resulting diffraction rings from microcrystalline regions

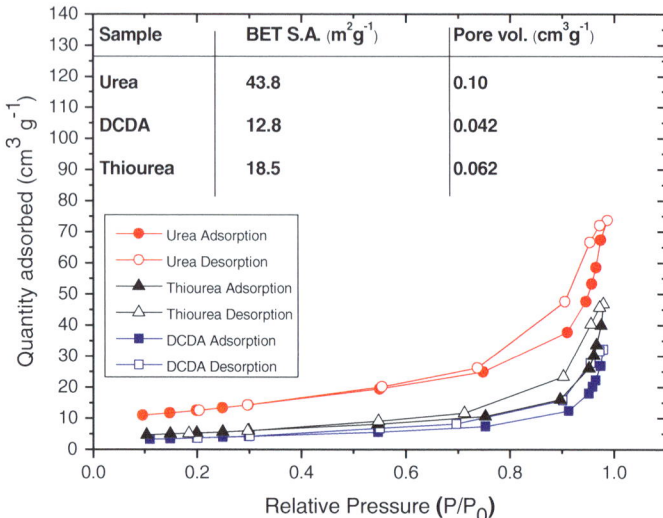

Fig. 4.6 BET nitrogen adsorption/desorption isotherms. All samples were pre-prepared by calcination at 600 °C, for 4 h, with a ramp rate of 5 °C/min

indicates pockets of polycrystallinity for the sample prepared using DCDA. Selective Area Electron Diffraction (SAED) in Fig. 4.5d shows diffraction ring patterns, one of which corresponds to 3.2 Å, the graphitic layer distance of g-C_3N_4. Graphitic carbon nitride synthesised from DCDA appears to be the densest structure, whilst the structure of urea-derived g-C_3N_4 appears more open and porous. This corresponds with the XRD patterns, and also reflects the BET surface areas of the samples (Fig. 4.6). Among the three samples, the urea-derived g-C_3N_4 is the least dense, exhibiting the largest surface area and also greatest average pore volume, followed by thiourea-derived g-C_3N_4, and then DCDA-derived g-C_3N_4. All isotherms can be classified as type V, demonstrating significant hysteresis and a plateau at $P/P_0 < 0.3$.

In order to understand the nature of the low density sample prepared from urea, thermal gravimetric analysis—differential scanning calorimetry—mass spectroscopy (TGA-DSC-MS) was undertaken to analyse the sample synthesis route, and the g-C_3N_4 prepared from thiourea was also analysed for comparison. Interestingly, new phenomena were observed; the origin of the porosity in g-C_3N_4 synthesised from urea and thiourea (Figs. 4.7 and 4.8).

According to a previous report, the condensation and polymerisation of cyanamide and dicyandiamide precursors occur with the gradual loss of NH_3, while thiourea could go through additional steps of losing not only NH_3, but potentially also CS_2 [19]. The TGA-MS results confirm this hypothesis, with both CS_2 and CH_4N_2 being detected by MS, in addition to NH_3 and thiocyanic acid. This accounts for the slight difference in surface area between DCDA and thiourea; both CS_2 and CH_4N_2 being trapped in the structure beyond their boiling points

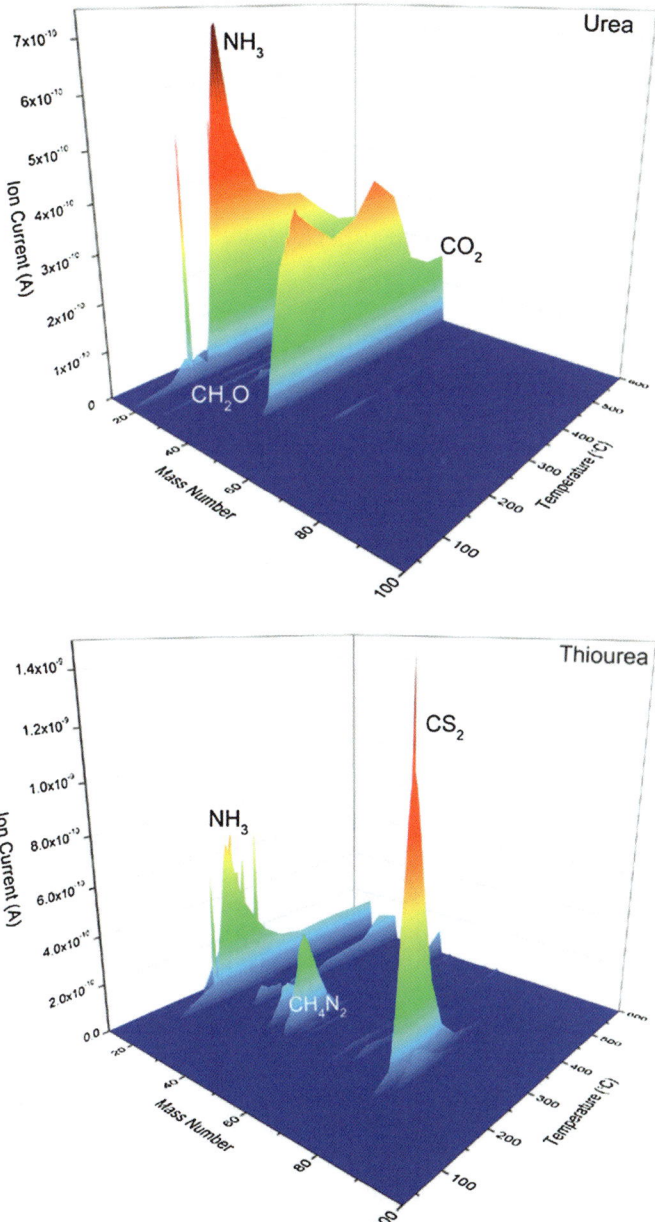

Fig. 4.7   TGA-DSC-MS data from g-C$_3$N$_4$ synthesised from urea and thiourea

and eventually ceasing to evolve at 192 and 310 °C as recorded by MS. Only trace amounts of CS$_2$ and CH$_4$N$_2$ remain at the point of polymerisation (500 °C). Crucially, urea is different from other precursors because of the presence of

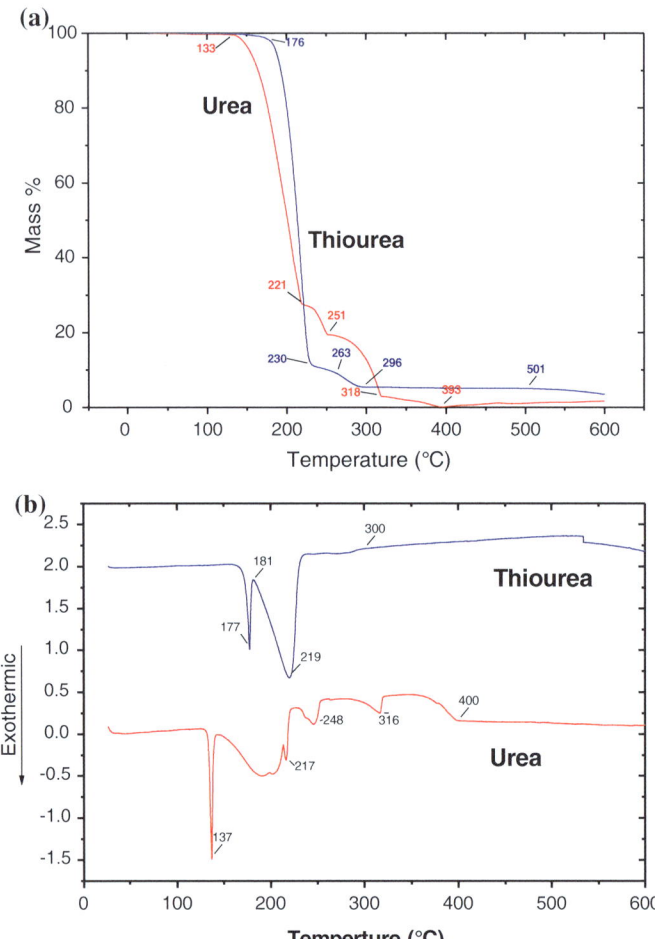

Fig. 4.8 TGA **a** and DSC **b** analysis of urea and thiourea

oxygen; CO_2 is initially detected at 150 °C, and continues to evolve slowly along
with NH_3, which is in a relatively high amount throughout the analysis. Since the
polymerisation of heptazine to melem happens around 389 °C [2], CO_2 can thus
aid the formation of pores; graphitic layers will grow around the trapped gas, as
shown in the fabrication of high porosity activated carbon [28]. In contrast to previous reports which propose other mechanisms behind the large surface area of
urea based carbon nitrides [29], it is proven here that CO_2 is crucial for the formation of the largest pores and the highest surface area.

Formaldehyde (CH_2O) was detected with a peak starting at 230 °C and continu-
ing throughout the heating process, possibly as a result of hydrogenation of CO
or CO_2. This would suggest that hydrogen is continuously removed from nitrogen in urea when reacting with CO to form CH_2O, which is then detected by the

MS; thereby enabling free nitrogen to covalently bond, forming melamine from precursory intermediate carbon, pinning the electron lone pair to nitrogen and leading to a high degree of polymerisation. This does not take place in the synthesis of C_3N_4 using DCDA/thiourea and is in agreement with the following discussion about the differences in surface chemistry between the three samples.

Graphitic carbon nitride, synthesised from different precursors (600, 5 °C/min ramp rate), was tested for hydrogen evolution in an aqueous sacrificial solution containing triethanolamine (TEOA) at room temperature and atmospheric pressure, akin to other reports [2, 8]. The fully optimised results are shown in Fig. 4.9 and further summarised in Table 4.2. As shown in Fig. 4.9a, b, the urea-derived g-C_3N_4 exhibits superior hydrogen evolution in comparison to either the widely used DCDA or thiourea derived g-C_3N_4 under both full arc and visible light irradiation. The urea-derived g-C_3N_4 evolves hydrogen at ca. 20,000 ¼ mol h^{-1} g^{-1}, 15 times faster than the DCDA derived and 8 times higher than thiourea-derived g-C_3N_4. This is reflected in the TON (over platinum cocatalyst); urea-derived g-C_3N_4 has a TON of 641.1, much higher than other samples.

Even under visible light irradiation ($\lambda \geq 395$ nm, Fig. 4.9b), the urea-derived g-C_3N_4 evolves H_2 at 3300 μmol h^{-1} g^{-1}, nearly 10 times faster than the DCDA-derived g-C_3N_4 at 300 μmol h^{-1} g^{-1} and nearly 7 times greater than thiourea-derived g-C_3N_4 at 500 μmol h^{-1} g^{-1}. One may conclude that the activity difference is due to surface area; however the urea-derived g-C_3N_4 only has a specific surface area (SSA) 3.4 times greater than DCDA derived g-C_3N_4 and 2.4 times of the sample thiourea derived g-C_3N_4 (Fig. 4.6). Meanwhile, the activity is over 15 times and more than 8 times higher compared to the samples synthesised using DCDA and thiourea respectively. Interestingly, urea derived g-C_3N_4 calcined at 550 °C shows a surface area of 83.5 m^2 g^{-1}, double that of the sample calcined at 600 °C. Yet, the activity of the former is only a third of the latter (Fig. 4.10). Jun et al. prepared g-C_3N_4 at 550 °C (3 h) using urea as a precursor, and reported a hydrogen evolution rate (HER) of 625 μmol h^{-1} g^{-1}, which is similar to the sample prepared here, and thus used as a control [22]. The difference in activity is due to the shorter calcination time employed during sample preparation, which will affect the level of polymerisation of the compound, as shown by Zhang et al. [30]. Curiously, despite being able to absorb less light, the urea derived carbon nitride still outperforms both DCDA and thiourea derived counterparts (Table 4.2). Therefore, the activity cannot be directly attributed to either surface area or optical absorption.

It was found that 600 °C is the optimal calcination temperature for g-C_3N_4 when considering the correlation with hydrogen production (Fig. 4.9a). This can be explained because of the behaviour of graphitic carbon nitride with temperature. As seen in the TGA-DSC-MS analysis (Figs. 4.7 and 4.8), the urea-derived g-C_3N_4 continually undergoes loss of NH_3 and eventually begins to form heptazine units at 500 °C. However, at this temperature, some of the precursors still remain and full conversion to g-C_3N_4 has not been completed. As the temperature rises, polymerisation progresses, and the compound becomes more polymeric in nature; this is reflected in the activity, as it increases with temperature. However,

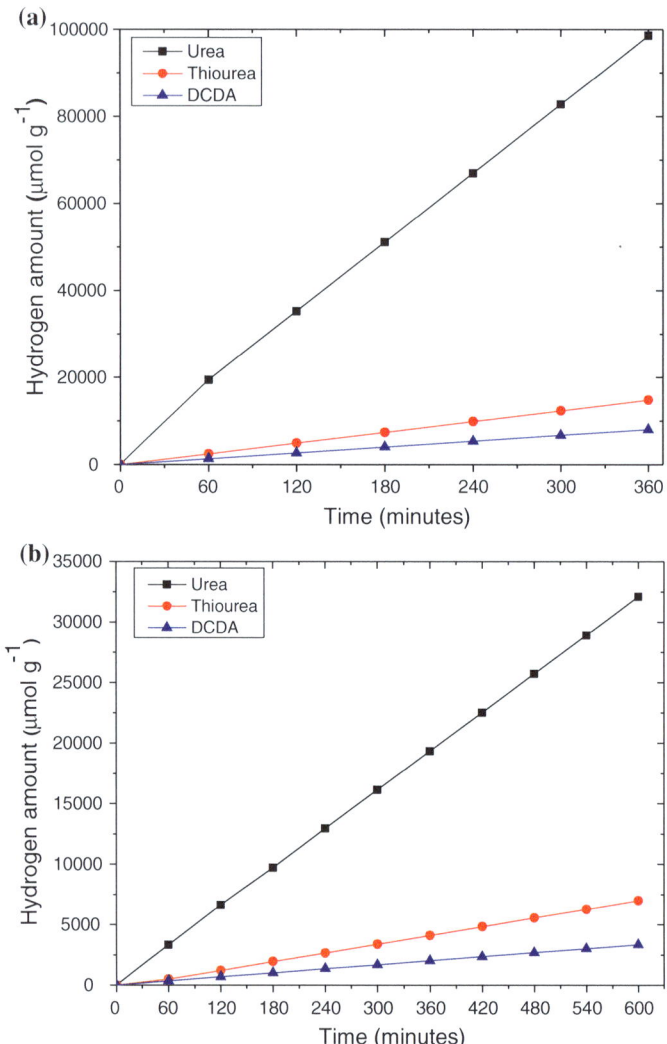

Fig. 4.9 Hydrogen evolution using a 300 W Xe lamp, 3 wt% Pt, TEOA as a hole scavenger; **a** Full arc, **b** λ ≥ 395 nm

Table 4.2 Summary of g-C_3N_4 properties synthesised from different precursors

Sample	HER rate (¼ mol h^{-1} g^{-1})	TON (6 h)	Band edge (nm)	SSA (m^2 g^{-1})
Urea	19,412	641.1	415	43.8
DCDA	1350	52.5	451	12.8
Thiourea	2470	96.4	453	18.5

H_2 evolution rate is using a 300 W Xe lamp, 3 wt% Pt, TEOA as a hole scavenger. All samples where synthesised at 600 °C in air. TON is calculated with respect to Pt catalyst

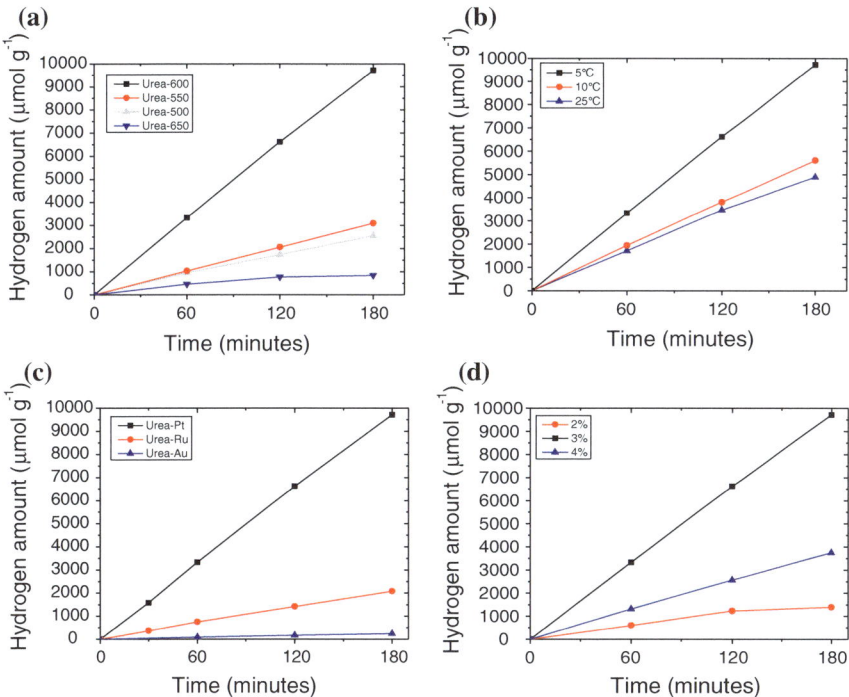

Fig. 4.10 Hydrogen production with various synthesis parameters of urea derived g-C$_3$N$_4$ ($\lambda \geq 395$ nm, TEOA hole scavenger, 300 W Xe lamp). **a** Urea based g-C$_3$N$_4$ synthesized at different temperatures, 4 h hold time (°C). **b** Urea synthesized at 600 °C using different ramping temperatures (°C/min). **c** g-C$_3$N$_4$ (urea) with different cocatalysts, 3 wt%. **d** Weight percent of Pt in comparison to photocatalyst (urea, 600 °C)

600 °C is a critical point; if the temperature reaches 650 °C, then the graphitic structure begins to break down, as full condensation of the network proceeds. As this happens the structure will buckle, and thus the activity will dramatically drop; the electronic structure is distorted and the compound ceases to be an effective photocatalyst, as seen elsewhere [4]. Non-polymeric carbon nitride, i.e. a melem unit, is not a photocatalyst, whilst g-C$_3$N$_4$ is. Therefore it can be hypothesised that a less polymeric g-C$_3$N$_4$, i.e. more breakages in the structure, essentially becomes less photocatalytic, and more inert.

As discussed later, the degree of polymerisation of g-C$_3$N$_4$ affects the protonation status, and thus the compound s ability to function as an effective photocatalyst. The ramp rate of the reaction is also important; (Fig. 4.10b) illustrates that a ramp rate of 5 °C is clearly superior to 10 and 25 °C, which appears to be a crucial step in enhancing photoactivity. From our experimentation, if the ramp rate is too fast, g-C$_3$N$_4$ has a tendency to sublimate, notably decreasing product yield. Subsequently, since condensation in this case is through the loss of ammonia groups, logic would dictate that sufficient time is necessary, at each polymerisation step, for NH_x and CH_2O to be released. If the reaction proceeds slowly, the network has the necessary time to form and proceed through each step, and gradually lose protons from within the structure, and polymerise into an effective photocatalyst.

Different cocatalysts were also deposited onto the surface of urea-derived g-C_3N_4; platinum was found to be the most effective at reducing protons to hydrogen because of the very small overpotential it provides (Fig. 4.10c). The weighting was also key; 3 % was found to be the optimum loading, with both 2 and 4 %; providing insufficiently and overly populated active sites, respectively, as reported on other photocatalysts loaded with noble metals (Fig. 4.10d) [31]. Large platinum loadings could also prevent light adsorption and halt redox reactions. It can also be argued that the weighting could influence size of the platinum particles deposited on the surface, which could then influence the catalytic activity of the system. However, there are no definitive/reliable studies to date, yet this could be a considerable future project if it was possible to produce platinum nanoparticles with very small size distributions.

X-ray photoelectron spectroscopy (XPS) studies were undertaken to accurately determine the specific bonding and structure of the samples. In all samples the typical C1s and N1s peaks are observed, similar to other reports [29, 32]. A residual O1s peak is also noted, which is likely due to calcination in air. XPS spectra from the three samples (urea, thiourea, DCDA) are shown in Fig. 4.11 (N1s XPS) and Fig. 4.12 (C1s and O1s XPS spectra). The O1s level (Fig. 4.12) is composed of two convoluted peaks at 532 and 534 eV, respectively, which are assigned to N-C-O (oxygen in the bulk) and adsorbed oxygen [33], presumably due to air contamination from calcination.

The C1s spectra show bonding of C-C, C-N-C and a trace C-O located at 285.1, 288.2, and 289.2 eV, respectively. The spectra of N1s can be fitted to elucidate 4 separate signals, and provides a better idea of the bonding structure, since carbon spectra are susceptible to contamination. The N1s core levels at 398.7, 399.7 and 400.9 eV correspond to sp^2 C-N-C, sp^3 H-N-[C]$_3$ and C-NH_x (amino functional groups) respectively [33]. The weak peak at 404.4 eV can be attributed to terminal nitrate groups, charging effects, or excitations [19, 32]. Hybridized sp^3 nitrogen not only has three chemical bonds to carbon, but because of hybridization, can also possess bonded hydrogen perpendicular to the direction of the graphitic layer.

The atomic ratio of various nitrogen bonding configurations in all the three samples were calculated based on XPS area mapping (3×3), using a point-to-point distance of 0.7 mm, in order to give an average, homogenized data set. Figure 4.13 and Table 4.3 shows the distinctive trend in bonding ratios versus activity between samples. The ratio of sp^2 C-N-C to the sum of sp^3 H-N-[C]$_3$ and C-NH_x bonds (where the latter represents the total amount of protons) is 2.83 in urea, 2.7 for thiourea and only 2.31 in DCDA. As it is part of the heptazine ring, linked by a double and a single bond to two opposing carbons, the sp^2 bonded nitrogen is the principle participant that contributes to band gap absorption and therefore is an incredibly important part of the structure. Both hybridized sp^3 nitrogen and surface functional amino groups (C-NH_x) are also key features when

Fig. 4.11 XPS spectra from g-C₃N₄, for urea, DCDA and thiourea derived samples

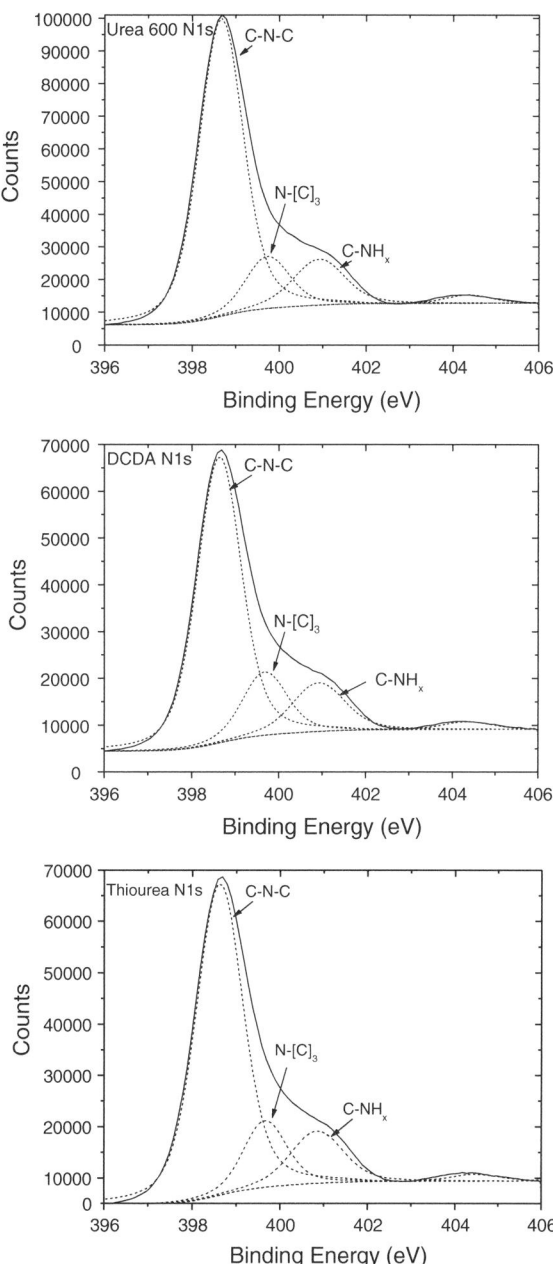

considering bulk and surface properties. Along with C-NH$_x$ bonding, graphitic carbon nitride possesses a positively charged, slightly acidic surface, confirmed by zeta potential measurements (Fig. 4.14). Elemental analysis (EA) further

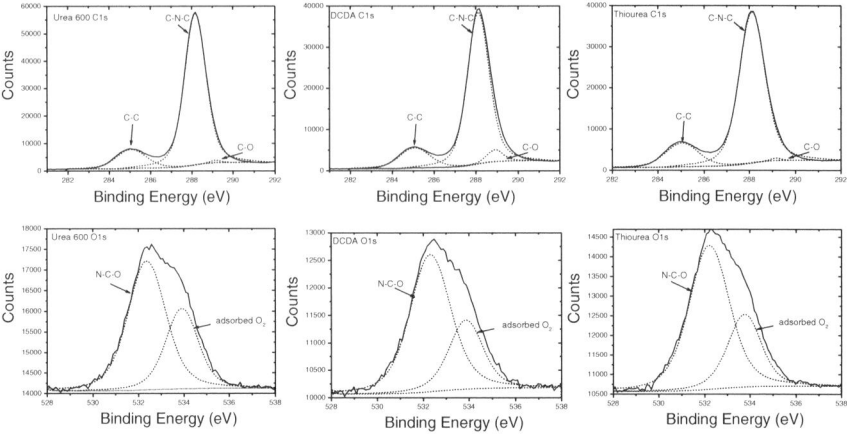

Fig. 4.12 C1s and O1s XPS spectra of g-C₃N₄ synthesised from different precursors

Fig. 4.13 Ratios of bonds within the N1s core level peak in different samples and their comparisons to the HER under visible light ($\lambda \geq 395$ nm), indicating decreasing proton concentration dramatically increases photocatalytic activity

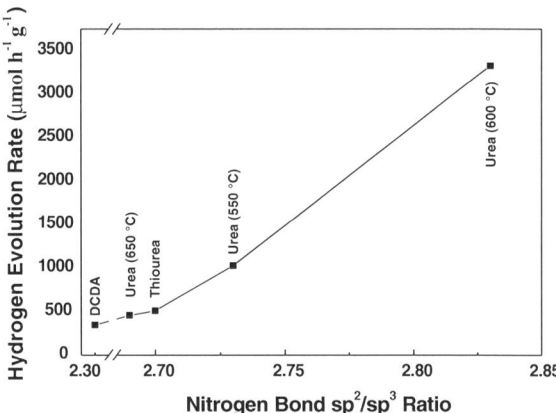

confirms the trend shown in Fig. 4.13; as bulk H (at.%) increases, the HER per SSA (mol m^{-2} h^{-1}) decreases (Fig. 4.15). Combining with XPS analysis, one can conclude that a lower proton concentration leads to a larger HER.

Zeta potential measurements of the three samples show a similar trend to XPS at the operating pH (9.75); the urea derived sample has the smallest zeta potential (i.e. smallest level of surface protonation), followed by the thiourea-derived and ultimately the DCDA-derived sample. Thus there is an evident correlation herein between activity and surface protonation; the lower the positive charge on the surface, the higher the activity. The negative charge seen at the slipping plane clearly shows the actual surface is positive; thus in a solution, attracts OH$^-$ groups.

EA results showed that DCDA, thiourea and urea (550, 600 and 650 °C) derived g-C₃N₄ possessed 17.36, 18.43, 18.72, 15.93, 15.45 atom% of hydrogen respectively (Fig. 4.15). Protons/hydrogen in g-C₃N₄ is present at sp^3 bound

Table 4.3 Ratios of bonds within the N1s core level peak in different samples and their comparisons to the hydrogen evolution activity

Sample	Ratio	
	sp^2 to sum of sp^3 and C-NH	Hydrogen evolution activity under visible light (μmol h^{-1} g^{-1})
Urea (600 °C)	2.83	3327
Urea (550 °C)	2.73	1027
Thiourea	2.70	508
Urea (650 °C)	2.69	457
DCDA	2.31	345

Fig. 4.14 Zeta potential of g-C$_3$N$_4$ synthesised from different precursors (600 °C)

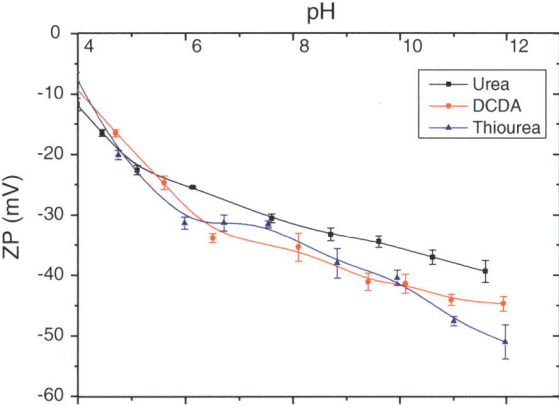

nitrogen sites (H-N-[C]$_3$), and as amino functional groups (C-NH$_x$). Proton/hydrogen atomic% represents bulk protonation status. It was found that the specific surface area (SSA, m^2 g^{-1}) normalised hydrogen evolution rate (HER, μmol m^{-2} h^{-1})—indicating the active sites per unit area for hydrogen production—is in accordance with the measured hydrogen content. Note that g-C$_3$N$_4$ starts to decomposed above 600 °C; urea (650 °C) derived g-C$_3$N$_4$ is therefore not included in the above comparison. This phenomenon further confirms the relationship between hydrogen evolution and proton concentration.

The increase in protons (from both sp^3 nitrogen, and C-NH$_x$) most likely stems from the route of condensation, and even though all three samples were synthesized using identical conditions, the extent of polymerization varies due to different precursors, which is consistent with the trends shown in XRD and FTIR measurements (Figs. 4.1 and 4.2). In particular, only urea-derived g-C$_3$N$_4$ loses hydrogen in the form of formaldehyde, due to the presence of oxygen in the precursor (TGA-DSC-MS; Figs. 4.7 and 4.8). This trend is not just between different precursors, in fact, urea-derived g-C$_3$N$_4$ synthesized at different temperatures (550, 650 °C) also follow suit (Fig. 4.13 and Table 4.3); as protonation increases, activity decreases. Therefore, both the precursors and synthesis parameters can

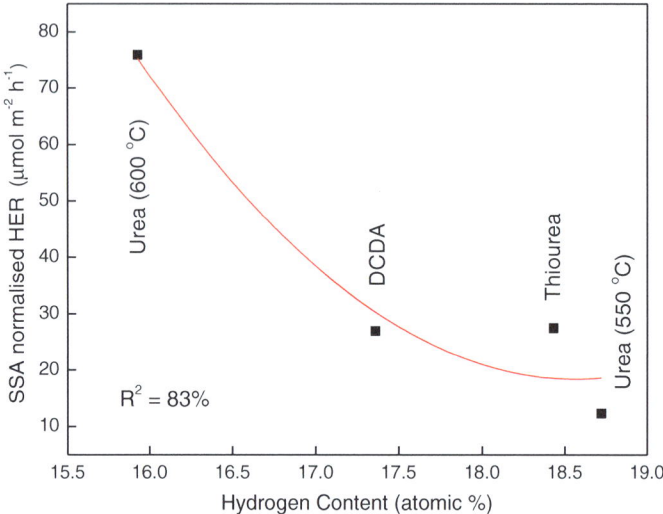

Fig. 4.15 Correlation between specific surface area normalised hydrogen evolution rate and bulk hydrogen content measured by elemental analysis

control the protonation and polymerisation, leading to varying activity. A small change in preparation method can have a huge impact on photocatalytic hydrogen production (such as from 550 to 600 °C and then to 650 °C; Fig. 4.10a and ramp rate changes Fig. 4.10b). Previous studies on the protonation status of graphitic carbon nitride have shown that the absorption profile can be shifted by treating compound with HCl, and thus protonating the extrinsic surface structure [30]. Therefore it is reasonable to assume that by altering the synthetic parameters, it is plausible that the intrinsic structure has changed during polymerisation, which could explain the change in absorption (Fig. 4.3) and hydrogen evolution rate from water (Fig. 4.9).

In order to futher determine why polymerisation and protonation status influences H_2 production rates, protonation was modelled using DFT simulations with periodic supercells. TDDFT simulations were also performed on smaller cluster models in order to determine the effects of hydrogen on excited state properties.

The density of states (DOS) is shown in Fig. 4.16c. It can be clearly seen that the CBE of the protonated system is shifted down in energy (towards more positive values with respect to the normal hydrogen electrode (NHE)) by 0.34 eV. This significantly modifies electrochemical properties as it provides a lower overpotential for reduction reactions, also shown in the UV-Vis absorption spectra (Fig. 4.3c). The reason behind the drop in the position of the CBE can be clearly seen in the site-decomposed DOS, Fig. 4.17. The effects of protonation on excited state properties were also calculated. It was established that the lowest energy vibrationally stable structure is away from a planar configuration. The onset of structure involves strong distortions of all three heptazine rings optical absorption

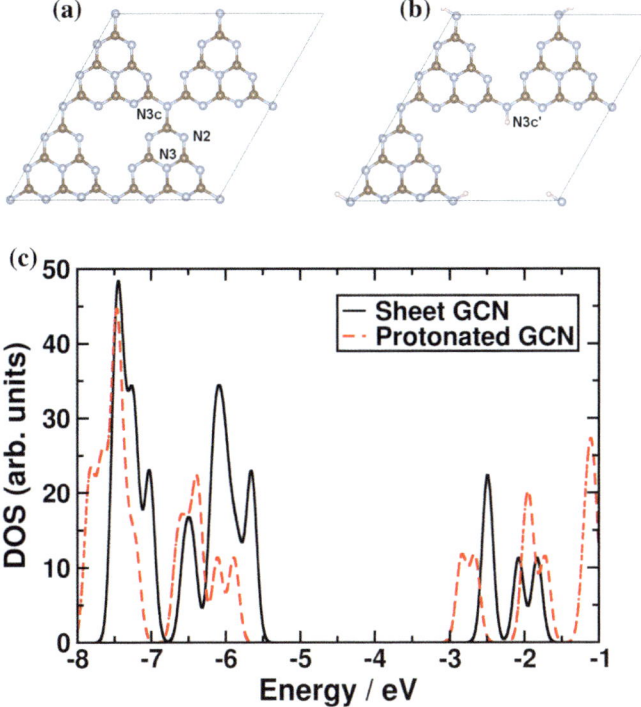

Fig. 4.16 Geometric and electronic structure of single sheet g-C_3N_4. **a** Supercell model of sheet carbon nitride, **b** supercell model of protonated carbon nitride. Nitrogen is denoted by *blue* spheres, carbon *brown* spheres, and hydrogen *light pink* spheres. **c** Total density of states for sheet carbon nitride (*black line*) and protonated carbon nitride (*red line*). Energy is with respect to the zero of the sheet carbon nitride simulation. The DOS of the protonated carbon nitride has been shifted so that the corresponding zero points align

Fig. 4.17 Total and site-decomposed DOS for protonated carbon nitride model. The strong contribution of the carbon atoms adjacent to the N3c site at the CBE is apparent, as the six atoms contribute almost as much to the DOS as the other twelve carbon atoms in the supercell

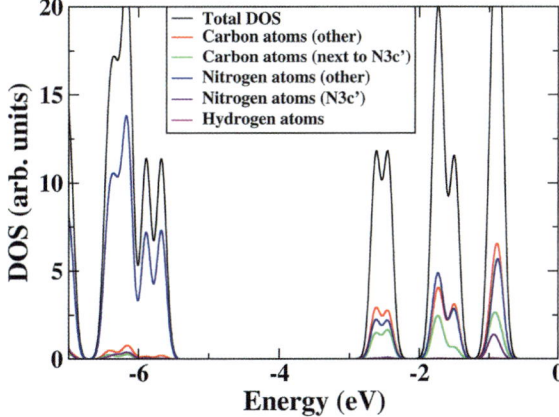

of protonated g-C$_3$N$_4$ occurs at a lower energy (more positive with respect to the NHE) than for un-protonated g-C$_3$N$_4$. Indeed, two absorption peaks of the C$_{18}$N$_{28}$H$_{13}$ model occur at lower energies than the initial absorption peak of the C$_{18}$N$_{28}$H$_{12}$ model, in qualitative agreement with the DFT DOS in Fig. 4.16. This verifies the DFT-based electronic structure analysis with TDDFT.

The distribution of the lowest energy exciton for both carbon nitride models was also plotted. The results are shown in Fig. 4.18b, c. For un-protonated g-C$_3$N$_4$ the exciton is distributed homogeneously over the cluster, with transitions from occupied N p$_z$ orbitals to empty C p$_z$ orbitals. For protonated g-C$_3$N$_4$, the exciton is more heterogeneous, with the photohole on the protonated-heptazine ring and the photoelectron distributed evenly on the other two heptazine rings. Although there is a better spatial separation between photohole and photoelectron, both charge carriers are more localized around the central N3 site, and thus are not as available to participate in the photochemical reactions. Moreover, this localisation will act to increase the exciton recombination rate, hindering the efficient utilisation of charge carriers.

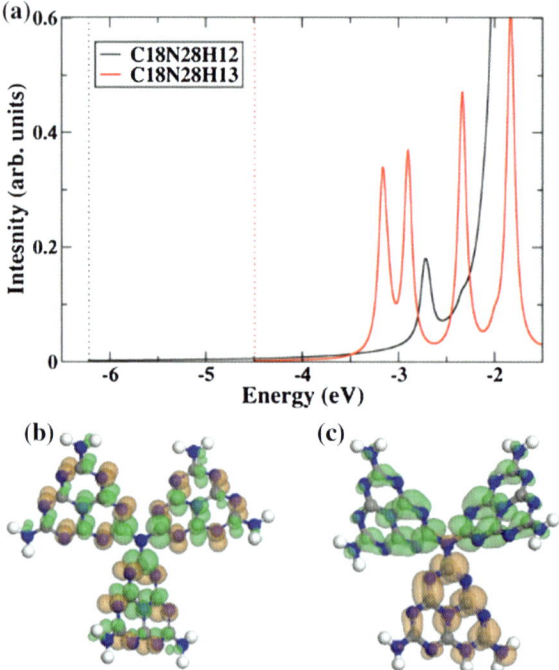

Fig. 4.18 Excited state properties of bare (sheet) and protonated carbon nitride; **a** optical properties of C$_{18}$N$_{28}$H$_{12}$ (*black line*) and C$_{18}$N$_{28}$H$_{13}$ (*red line*) clusters, the x-axis is labeled with respect to the vacuum level (E = 0) while dashed lines indicate the positions of the HOMO states of the two clusters, **b** lowest energy exciton of C$_{18}$N$_{28}$H$_{12}$ cluster, **c** lowest energy exciton of C$_{18}$N$_{28}$H$_{13}$ cluster. Orange isosurface indicates distribution of photohole upon photoexcitation, green isosurface indicates distribution of photoelectron upon photoexcitation. Isosurfaces were plotted at 0.005 |e|/Å^3. Blue spheres denote hydrogen atoms, grey spheres denote carbon atoms, white spheres denote hydrogen atoms

A relatively small change in protonation has a two-pronged detrimental effect on the reduction ability of g-C_3N_4. Protonation significantly reduces the reduction potential and also localizes excitons around a central nitrogen N3 site, hindering migration to active sites. Here, protonation is essentially controlled through the degree of polymerisation, but also coupled with the degree of condensation. Graphitic carbon nitride, if "under-polymerised" (at low temperatures), has incomplete heptazine coupling, which results in excess hydrogen passivating N3c' nitrogen sites, hampering activity. If "over-polymerised" (at temperatures ca. 650 °C), the structure of g-C_3N_4 tends to overly condense into buckled multi-layered crystals of reduced surface area and hence reduced density of active sites, adversely effecting photocatalytic ability. Moreover, the buckling also distorts the sp^2 planar geometry, leading to charge trapping states at nitrogen sites and hence reduced activity.

As mentioned in Sect. 1.3.2, apart from a photocatalyst needing to be cheap and robust, it must exhibit a high quantum yield for hydrogen production from water under visible light if it is even to be considered commercially viable. Compounds that traditionally exhibit high efficiencies either suffer from instability (e.g. sulphides [34, 35]) or are made of relatively expensive metals (e.g. GaAs-GaInP$_2^{68}$). It is shown in Fig. 4.19a the cheap and stable urea-derived g-C_3N_4 has a peak internal quantum yield of 28.4 % at 365 nm. Even under visible light irradiation at $\lambda = 400$ nm, the quantum yield is 26.5 %, nearly an order of magnitude greater than the highest reported (3.75 % at 420 nm[7f], obtained via liquid exfoliation). In order to assure the reliability of our measurement, we repeated a benchmark cyanamide derived g-C_3N_4 as a reference which showed comparable activity (small difference is due to a 395 nm long pass filter used instead of a 420 nm filter).

As proposed previously, the huge enhancement in hydrogen evolution rate of our urea-derived sample (3327.5 vs. 142.3 μmol h^{-1} g^{-1}) can be attributed to a lower protonation status and condensation state. Very recently, a facile synthesis method for three-dimensional porous g-C_3N_4 was introduced by using the aggregates of melamine and cyanuric acid (MCA) co-crystals in dimethyl sulfoxide (DMSO, sample denoted MCADMSO) as precursors [22]. It was reported the quantum yield of g-C_3N_4MCADMSO at $\lambda = 420$ nm was 2.3 % (Table 4.4), much higher than the melamine derived bulk g-C_3N_4 (0.26 %) under the same conditions. This study was repeated, and as shown in Fig. 4.20, very similar morphologies and optical properties compared with the literature were measured. A similar quantum yield of 3.1 % at $\lambda = 400$ nm (Table 4.4) was obtained (difference is due to the wavelength of band-pass filter). Since MCA_DMSO is another oxygen-containing precursor, the rise in the quantum yield compared with a melamine sample further supports the proposed protonation mechanism. In addition, a theory as to why the optimised urea derived g-C_3N_4 demonstrates a tenfold increase in IQY yield at 400 nm, in comparison to MCA_DMSO g-C_3N_4, is because of the much higher oxygen concentration in the urea precursor, which helps to passivate protonation sites and polymerise g-C_3N_4 without structural instability/buckling.

The stability of the optimised photocatalyst was also tested in an extended experiment, as illustrated in Fig. 4.19b. The high activity is reproducible and the

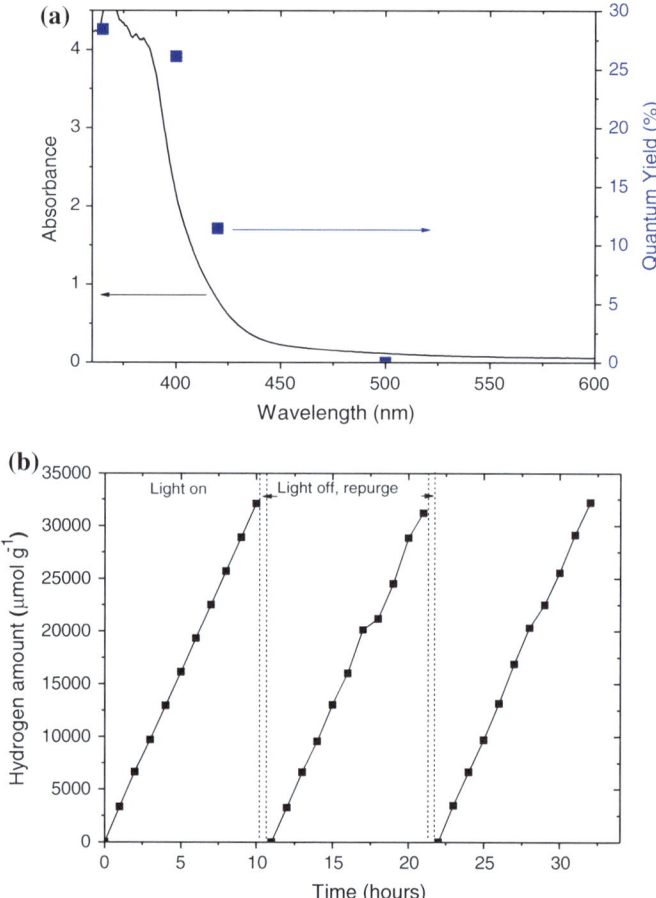

Fig. 4.19 a Quantum yield of urea based g-C$_3$N$_4$; using band pass filters at specific wavelengths (*black line* represents absorbance, internal quantum yield is shown by *blue points*). **b** Stability test of the urea derived g-C$_3$N$_4$ under visible light irradiation ($\lambda \geq 395$ nm)

Table 4.4 Percentage breakdown of different bonds within the N1s spectrum

	N1s	Atomic % of bond (sample calcined at 600 °C, 4 h, 5 °C/min ramp rate)		
Bond	Binding energy (eV)	Urea	DCDA	Thiourea
C-N-C (sp^2)	398.7	72.3	68.4	71.8
N-[C]$_3$ (sp^3)	399.7	12.6	15.6	13.6
C-NH$_x$	400.9	13.0	14.0	13.0
NO$_2$ terminal	404.4	2.1	2.1	1.7

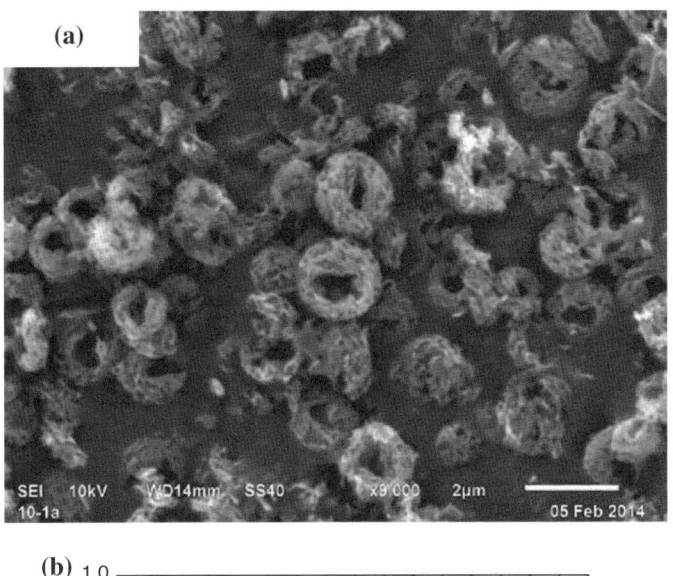

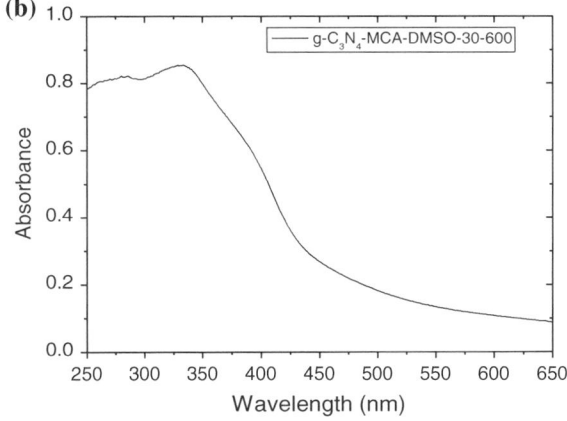

Fig. 4.20   **a** SEM micrograph of g-C$_3$N$_4$-MCA-DMSO-30-600. **b** UV-Vis absorbance spectra of g-C$_3$N$_4$-MCA-DMSO-30-600

material shows excellent stability within a 30 h run. Attributes for a high quantum yield commonly include good absorption, efficient charge separation and rapid carrier transfer to the surface for redox reactions. Even though the overall band gap of urea based carbon nitride is larger than that of both DCDA and thiourea derived carbon nitrides, it produces more hydrogen, therefore the bulk absorption and band gap magnitude are not the determining factor in overall activity. Due to the fewer protons in g-C$_3$N$_4$ (urea), the bandgap and therefore the overpotential is larger, which also results in better separation of charge, and consequently causes better migration of charger carriers to active sites. The trend of the proton

Table 4.5 Comparison of typical g-C_3N_4 photocatalysts reported for hydrogen production and the corresponding quantum yields

Photocatalyst[a]	Band gap (eV)	HER rate under visible light (μmol h^{-1} g^{-1})[b]	IQY, % (band pass filter, nm)	Reference
g-C_3N_4 (cyanamide)	2.7	106	ca. 0.1 (420–460 nm)	[7]
mpg-C_3N_4	2.7	1490	N/A	[5]
g-C_3N_4 (S-doped)	2.85	800	N/A	[3]
g-C_3N_4 nanosheets	2.35	1860	3.7 (420 nm)	[8]
g-C_3N_4 (MCA-DMSO)	2.83	N/A	2.3 (420 nm)	[22]
Our g-C_3N_4 (cyanamide)	2.7	142	N/A	This work
Our g-C_3N_4 (MCA-DMSO)	2.8	261	3.1 (400 nm)	This work
Our g-C_3N_4 (urea)	2.85	3327	26.5 (400 nm)	This work

[a]All photocatalysts are loaded with 3 wt% Pt co-catalysts except g-C_3N_4 (S-doped) which is mixed with 6 wt% Pt co-catalyst
[b]HER rates are measured either by using long pass ($\lambda \geq 420$ nm) or ($\lambda \geq 395$ nm) such as for our g-C_3N_4 (cyanamide) and our g-C_3N_4 (urea), therefore the small difference is due to the long pass filters used

concentration and the trend of polymerisation for the sample set are in very good agreement with the corresponding activity, which contributes to charge transport. Therefore for the first time, it has been demonstrated that the protonation and degree of polymerisation determines hydrogen evolution rates from water, with surface area playing a minor role [1] (Table 4.5).

4.4 Conclusions

To conclude, a novel strategy for synthesising graphitic carbon nitride has been demonstrated, by thermally decomposing urea at appropriate temperature (e.g. 600 °C), ramp rate (5 °C) and hold time (4 h). Several different precursors were used, including urea, thiourea, and DCDA. XRD patterns exhibit varying peak intensities indicating the most crystallised/condensed sample is DCDA, and urea the least, as a more intense peak implies that there is more regular repetitions between graphitic layers. UV-Vis spectra shows the sample synthesised from thiourea shows the largest absorption, with an edge around 455 nm. The DCDA derived sample absorbs up to ca. 450 nm, whilst the sample fabricated using urea exhibited a visible blue shift, absorbing light up to approximately 420 nm. All the samples have a small absorption tail, due to some sub band gap states from n-π^* transitions. TGA-DSC-MS data shows that as urea based g-C_3N_4 is thermally decomposed, CO_2 is given off in large amounts, and up to the polymerisation point of heptazine units (over 500 °C). The presence of CO_2 is crucial for the formation of the largest pores and the highest surface area—and is not present in either the decomposition of thiourea or DCDA. Furthermore, formaldehyde is also given off throughout decomposition, and continually removes hydrogen from the structure, which helps to promote polymerisation.

The internal quantum yield in the visible region by the optimised sample is 26.5 %, nearly an order of magnitude higher than reported previously. Under full arc irradiation, the optimized g-C_3N_4 can be tailored to produce ca. 20,000 μmol g^{-1} of hydrogen per hour from water. The optimized g-C_3N_4 photocatalyst is very stable, exhibiting a near linear profile of H_2 production from water for 30 h and resulting in a TON of over 641 in terms of Pt cocatalyst after a 6 h test under 300 W Xe lamp irradiation. Data from XPS, FTIR, zeta potential and XRD show that both protonation status and degree of polymerisation can influence the g-C_3N_4 hydrogen evolution rate.

Specifically, from the N1 s core XPS data, we observe the ratio of sp^2 C-N-C to the sum of sp^3 H-N-[C]$_3$ and C-NH$_x$ bonds (where the latter represents the total amount of protons) is 2.83 in urea, 2.7 for thiourea and only 2.31 in DCDA. As it is part of the heptazine ring, linked by a double and a single bond to two opposing carbons, the sp^2 bonded nitrogen is the principle participant that contributes to band gap absorption and therefore is an incredibly important part of the structure. Both hybridized sp^3 nitrogen and surface functional amino groups (C-NH$_x$) are also key features when considering bulk and surface properties. Along with C-NH$_x$ bonding, graphitic carbon nitride possesses a positively charged, slightly acidic surface, confirmed by zeta potential measurements. Elemental analysis (EA) further confirms the trend; as bulk H (at.%) increases, the HER per SSA (mol m^{-2} h^{-1}) decreases. Combining with XPS analysis, one can conclude that a lower proton concentration leads to a larger HER.

Two different lines of computational evidence (DFT and TDDFT) have shown that the larger HER is a direct result of a decrease in surface protonation status. This is because of a shift in the position of the conduction band edge occurs, thus increasing the overpotential for reduction reactions at the surface. A larger overpotential allows reduction reactions to proceed more readily, despite increasing the band gap. Furthermore, there is a significant shift in the CBE position even for limited proton concentrations that are less than experimental values (6.5 % atomic H for DFT, and 1.6 % atomic H for TDDFT, versus 15.5 % atomic H evidenced by EA). Excess protonation also localizes photoelectrons at non-active N3 redox sites, effectively rendering the photoexcited charge useless.

The evidence contained in this chapter suggests that not only does urea based g-C_3N_4 exhibit a much larger efficiency in comparison to other samples, but also that this is a direct result of the band structure and surface chemistry. Therefore considering this enhancement and the energetic requirements of Z-Scheme systems, the next chapter will focus on the incorporation of g-C_3N_4 into a Z-Scheme system, a never before explored experiment due to the previously reported low quantum yields of carbon nitride at visible wavelengths.

References

1. Martin, D. J. et al (2014). Highly efficient H_2 evolution from water under visible light by structure-controlled graphitic carbon nitride. *Angewandte Chemie International Edition, 53*, 9240–9245.
2. Wang, X., et al. (2008). A metal-free polymeric photocatalyst for hydrogen production from water under visible light. *Nature Materials, 8*, 76–80.
3. Yue, B., Li, Q., Iwai, H., Kako, T., & Ye, J. (2011). Hydrogen production using zinc-doped carbon nitride catalyst irradiated with visible light. *Science and Technology of Advanced Materials, 12*, 034401.
4. Jorge, A. B., et al. (2013). H_2 and O_2 Evolution from water half-splitting reactions by graphitic carbon nitride materials. *The Journal of Physical Chemistry C, 117*, 7178–7185.
5. Ge, L., Han, C., Xiao, X., Guo, L. & Li, Y. (2014) Enhanced visible light photocatalytic hydrogen evolution of sulfur-doped polymeric $g\text{-}C_3N_4$ photocatalysts. *Materials Research Bulletin, 48*(4), 1447–1452.
6. Schwinghammer, K., et al. (2013). Triazine-based carbon nitrides for visible-light-driven hydrogen evolution. *Angewandte Chemie International Edition, 52*, 2435–2439.
7. Xiang, Q., Yu, J., & Jaroniec, M. (2011). Preparation and enhanced visible-light photocatalytic H_2-production activity of graphene/C_3N_4 composites. *The Journal of Physical Chemistry C, 115*, 7355–7363.
8. Wang, X., et al. (2009). Polymer semiconductors for artificial photosynthesis: Hydrogen evolution by mesoporous graphitic carbon nitride with visible light. *Journal of the American Chemical Society, 131*, 1680–1681.
9. Xu, J., Li, Y., Peng, S., Lu, G., & Li, S. (2013). Eosin Y-sensitized graphitic carbon nitride fabricated by heating urea for visible light photocatalytic hydrogen evolution: the effect of the pyrolysis temperature of urea. *Physical Chemistry Chemical Physics, 15*, 7657–7665.
10. Wang, Y., Zhang, J., Wang, X., Antonietti, M., & Li, H. (2010). Boron-and fluorine-containing mesoporous carbon nitride polymers: Metal-free catalysts for cyclohexane oxidation. *Angewandte Chemie International Edition, 49*, 3356–3359.
11. Liu, G., et al. (2010). Unique electronic structure induced high photoreactivity of sulfur-doped graphitic C_3N_4. *Journal of the American Chemical Society, 132*, 11642–11648.
12. Li, K., Martin, D. J., & Tang, J. (2011). Conversion of solar energy to fuels by inorganic heterogeneous systems. *Chinese Journal of Catalysis, 32*, 879–890.
13. Liebig, J. (1834). *Ueber Einige Stickstoff-Verbindungen. Annalen der Pharmacie, 10*, 1–47.
14. Franklin, E. C. (1922). The ammono carbonic acids. *Journal of the American Chemical Society, 44*, 486–509.
15. Pauling, L., & Sturdivant, J. H. (1937). The structure of cyameluric acid, hydromelonic acid and related substances. *Proceedings of the National Academy of Sciences, 23*, 615–620.
16. Lotsch, B. V., et al. (2007). Unmasking melon by a complementary approach employing electron diffraction, solid-state nmr spectroscopy, and theoretical calculations—structural characterization of a carbon nitride polymer. *Chemistry—A European Journal, 13*, 4969–4980.
17. Kroke, E., & Schwarz, M. (2004). Novel group 14 nitrides. *Coordination Chemistry Reviews, 248*, 493–532.
18. Kroke, E., et al. (2002). Tri-s-triazine derivatives. Part I. From trichloro-tri-s-triazine to graphitic C_3N_4 structures. *New Journal of Chemistry, 26*, 508–512.
19. Dong, F., Sun, Y., Wu, L., Fu, M., & Wu, Z. (2012). Facile transformation of low cost thiourea into nitrogen-rich graphitic carbon nitride nanocatalyst with high visible light photocatalytic performance. *Catalysis Science and Technology, 2*, 1332–1335.
20. Dong, F., et al. (2011). Efficient synthesis of polymeric $g\text{-}C_3N_4$ layered materials as novel efficient visible light driven photocatalysts. *Journal of Materials Chemistry, 21*, 15171–15174.

21. Bamwenda, G. R., Tsubota, S., Nakamura, T., & Haruta, M. (1995). Photoassisted hydrogen production from a water-ethanol solution: A comparison of activities of Au-TiO$_2$ and Pt-TiO$_2$. *Journal of Photochemistry and Photobiology A: Chemistry, 89*, 177–189.

22. Jun, Y.-S., et al. (2013). Three-dimensional macroscopic assemblies of low-dimensional carbon nitrides for enhanced hydrogen evolution. *Angewandte Chemie International Edition, 52*, 11083–11087.

23. Jürgens, B., et al. (2003). Melem (2,5,8-Triamino-tri-s-triazine), an important intermediate during condensation of melamine rings to graphitic carbon nitride: synthesis, structure determination by x-ray powder diffractometry, solid-state nmr, and theoretical studies. *Journal of the American Chemical Society, 125*, 10288–10300.

24. Deifallah, M., McMillan, P. F., & Cora, F. (2008). Electronic and structural properties of two-dimensional carbon nitride graphenes. *The Journal of Physical Chemistry C, 112*, 5447–5453.

25. Ferrari, A. C., & Robertson, J. (2000). Interpretation of Raman spectra of disordered and amorphous carbon. *Physical Review B, 61*, 14095–14107.

26. Ferrari, A. C., Rodil, S. E., & Robertson, J. (2003). Interpretation of infrared and Raman spectra of amorphous carbon nitrides. *Physical Review B, 67*, 155306.

27. Larkin, P. J., Makowski, M. P., & Colthup, N. B. (1999). The form of the normal modes of s-triazine: infrared and Raman spectral analysis and ab initio force field calculations. *Spectrochimica Acta Part A: Molecular and Biomolecular Spectroscopy, 55*, 1011–1020.

28. González, J. F., Román, S., González-García, C. M., Nabais, J. V., & Ortiz, A. L. (2009). Porosity development in activated carbons prepared from walnut shells by carbon dioxide or steam activation. *Industrial and Engineering Chemistry Research, 48*, 7474–7481.

29. Zhang, Y., Liu, J., Wu, G., & Chen, W. (2012). Porous graphitic carbon nitride synthesized via direct polymerization of urea for efficient sunlight-driven photocatalytic hydrogen production. *Nanoscale, 4*, 5300–5303.

30. Zhang, Y., Thomas, A., Antonietti, M., & Wang, X. (2008). Activation of carbon nitride solids by protonation: Morphology changes, enhanced ionic conductivity, and photoconduction experiments. *Journal of the American Chemical Society, 131*, 50–51.

31. Yamasita, D., Takata, T., Hara, M., Kondo, J. N., & Domen, K. (2004). Recent progress of visible-light-driven heterogeneous photocatalysts for overall water splitting. *Solid State Ionics, 172*, 591–595.

32. Thomas, A., et al. (2008). Graphitic carbon nitride materials: variation of structure and morphology and their use as metal-free catalysts. *Journal of Materials Chemistry, 18*, 4893–4908.

33. Li, J., et al. (2012). A facile approach to synthesize novel oxygen-doped g-C$_3$N$_4$ with superior visible-light photoreactivity. *Chemical Communications, 48*, 12017–12019.

34. Reber, J. F., & Meier, K. (1984). Photochemical production of hydrogen with zinc sulfide suspensions. *The Journal of Physical Chemistry, 88*, 5903–5913.

35. Matsumura, M., Saho, Y., & Tsubomura, H. (1983). Photocatalytic hydrogen production from solutions of sulfite using platinized cadmium sulfide powder. *The Journal of Physical Chemistry, 87*, 3807–3808.

Chapter 5
Novel Z-Scheme Overall Water Splitting Systems

Following the success of an extremely efficient photocatalyst for both O_2 and H_2 evolution under visible light, the research focus naturally moved to a demonstration system for pure water splitting by using urea derived g-C_3N_4 and tetrahedral Ag_3PO_4 crystals. It was found Ag_3PO_4 was also not able to function as an oxygen evolving photocatalyst in Z-scheme water splitting systems due to the incompatibility between photocatalyst and synthetic conditions. Ag_3PO_4 dissolves into the solution at the pH range (≤ 3.5) required for $Fe^{2+/3+}$ to be an effective redox mediator, and also partially reacts with H_2SO_4 to form Ag_2SO_4-Ag_3PO_4 as evidenced by XRD data. When NaI is used as a redox mediator, TEM and EDX results show that Ag_3PO_4 reacts with the iodide anion to form AgI, visibly changing the colour of the solution before illumination and also demonstrating no activity for overall water splitting.

Urea derived g-C_3N_4 is shown herein to actively participate in novel Z-scheme water splitting systems as a hydrogen evolving photocatalyst. Due to the robustness and favourable band position, g-C_3N_4 can be coupled with either I^-/IO_3^- or $Fe^{2+/3+}$ redox mediators, at any pH, and with Pt-loaded WO_3 or $BiVO_4$ respectively. The highest water splitting rates were achieved using a g-C_3N_4–NaI–WO_3 system, peaking at 36 and 18 μmol h^{-1} g^{-1} of hydrogen and oxygen respectively. The largest STH% was recorded at 0.1 %.

The redox sensitivity of the WO_3 (I/IO_3^-) water splitting system was demonstrated by testing WO_3 platelets, which have a smaller band gap than their commercial counterparts. The narrowing of the band gap, a result of morphological changes, was shown to have a dramatic negative effect upon the system; no water splitting was observed. Carbon nitride synthesized using precursors such as DCDA or thiourea do not give detectable activity for any Z-scheme water splitting reaction. It is postulated that due to the low efficiency of these carbon nitrides in comparison to urea-derived g-C_3N_4 for hydrogen evolution (as described in Chap. 4), overall water splitting rates are undetectable.

© Springer International Publishing Switzerland 2015   123
D.J. Martin, *Investigation into High Efficiency Visible Light Photocatalysts for Water Reduction and Oxidation*, Springer Theses,
DOI 10.1007/978-3-319-18488-3_5

5.1 Introduction

A single photocatalyst for overall water splitting has been explored for several decades but met with little success [1]. Nature splits water into O_2 and the H_2 equivalent species by a double excitation process including PSII and PSI, instead of a single excitation, in which the two half reactions are spatially separated and take place in PSII and PSI, Fig. 1.12 [2]. This overcomes the main problems of a singular photocatalytic water splitting system both kinetically and thermodynamically, as well described in previous reviews [3]. Inspired by natural photosynthesis, Bard proposed an analogy composed of two inorganic semiconductor photocatalysts in 1979 and recently there have been some productive systems based upon either Rh doped $SrTiO_3$ or TaON [4, 5].

The ideal artificial double excitation process is illustrated in Fig. 1.12. Briefly, a combination of two semiconductor photocatalysts, in which each photocatalyst is responsible for one half reaction and a soluble mediator helps electron transfer between the two photocatalysts so that an ideal cycle can be completed. A mediator is vital because it dramatically inhibits the fast unfavourable recombination of charge, analogous to the electron transport chain between PSII and PSI. Given such advantages of a double excitation process, there are many researchers working on either two photocatalysts each of which favours either H_2 or O_2 production, or a new mediator to efficiently transfer the charges between two photocatalysts. Thus a nature-inspired double excitation system composed of abundant elements and working efficiently for overall water splitting is highly sought-after.

5.2 Methodology

5.2.1 Photocatalytic Analysis

All reactions were carried out in a custom Pyrex® batch reactor cell, which was thoroughly purged with argon prior to radiation. Gas concentration analysis was performed using a GC (Varian 430-GC, TCD, argon carrier gas 99.999 %). Platinum, acting as a cocatalyst, was deposited onto g-C_3N_4 using an in situ photodeposition method. Stock solutions of deionised water and precursors (e.g. Pt: $H_2PtCl_6 \cdot (H_2O)_6$) were made beforehand, and simply added according to the required weight of metal (various wt% was tested previously, 3 wt% is optimum). The reactor was sealed, purged with Ar gas for 1 h, and then irradiated for 1 h under full arc irradiation using a 300 W Xe lamp (TrusTech PLS-SXE 300/300 UV). After 1 h, a measurement was taken to determine if hydrogen had been produced, and thus, if photodeposition had occurred correctly. The photocatalyst was then collected, washed three times, and dried overnight.

Platinum was deposited onto WO_3 using an impregnation method. WO_3 powder was put in an aqueous solution of $H_2PtCl_6 \cdot (H_2O)_6$, and heated until all water had evaporated. The powder was then heated at 500 °C for 1 h. After appropriate cocatalysts had been deposited, photocatalysts were immersed in an aqueous redox mediator solution, with different concentrations. The system was sealed, purged with argon gas for 1 h to remove all air in solution and headspace. A baseline measurement was taken, and then the reactor was irradiated with a 300 W Xe lamp (TrusTech PLS-SXE 300/300 UV). A 395 nm long pass filter was used to remove UV radiation (Comar Optics) where necessary.

Light intensity measured at 5 different points to obtain an average intensity. The solar to hydrogen efficiency was then calculated using the following formula; STH% = (Gibbs free energy of water × hydrogen production rate/energy density of incoming irradiation × area of light beam) × 100.

5.2.2 Synthesis Techniques

Highly efficient graphitic carbon nitride was synthesised using urea (Sigma-Aldrich 99.9 %). In a typical run, urea was placed in a lidded high quality alumina crucible, then put inside a muffle furnace and heated to 600 °C for several hours (5 °C min^{-1} ramp rate). The whole procedure was then carefully optimised to give the highest activity for H_2 production from water, as previously described in Chap. 4. Conventional photoactive monoclinic WO_3 was purchased directly from a supplier (Sigma-Aldrich) and further ground into a fine powder. $BiVO_4$ was synthesised using a method described by Kudo et al. [6]. Monoclinic WO_3 platelets were synthesised using a facile recipe tailored in house, based on a classical fabrication route to synthesise non-faceted monoclinic WO_3 [7]. The classical route uses $CaWO_4$ as the precursor, which is not soluble in water. Here $Na_2WO_4 \cdot 2H_2O$ is dissolved (1 g, Alfa Aesar 99.8 %) in 50 cm^3 DI water, after which 5 M HCl was added dropwise until a green H_2WO_4 precipitate appeared. This was then stirred for 24 h, washed with DI water, and filtered. The precipitate was then dried overnight and subsequently placed into a lidded alumina crucible, and calcined in a muffle furnace at 500 °C for 6 h, with a ramp rate of 10 °C min^{-1}. The WO_3 platelets were then ground into a fine powder.

5.3 Results and Discussion

5.3.1 Ag₃PO₄ Based Z-Scheme Water Splitting Systems

Attempts were made to utilise Ag_3PO_4 in Z-Scheme system, coupled with g-C_3N_4. Tetrahedral silver phosphate, synthesised using Method 'D' in Chap. 3 immersed first (before g-C_3N_4) in an aqueous 2 mM $Fe^{2+/3+}$ (starting with $FeCl_3$) redox mediator solution. H_2SO_4 was used to adjust the pH to 3, at which point Ag_3PO_4 completely dissolved into the solution, and thus the experiment was forced to stop. At pH 4, which is essentially the upper pH limit at which $Fe^{2+/3+}$ does not

Fig. 5.1 Powder X-ray diffraction data post water splitting in g-C$_3$N$_4$–Fe$^{2+/3+}$– Ag$_3$PO$_4$ system. Both main phase silver phosphate and sub phase silver sulphate are indicated by *red* and *blue* bars respectively

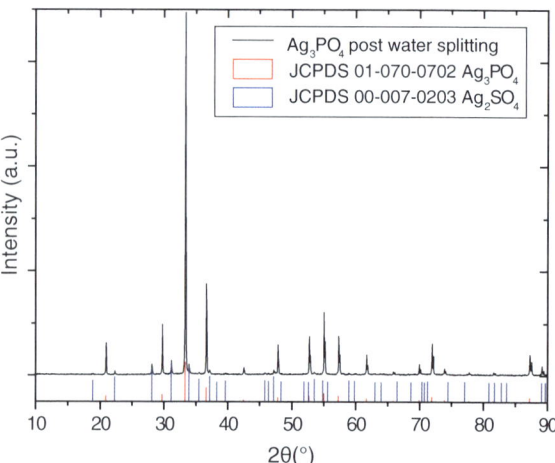

undergo hydrolysis, Ag$_3$PO$_4$ appeared stable in the solution. However, a slight discolouration of the suspension was cause for concern. The particles were removed from the reactor, washed, and centrifuged 3 times. After which, XRD results (Fig. 5.1) confirm that Ag$_3$PO$_4$ indeed did react with H$_2$SO$_4$ to form an Ag$_3$PO$_4$-Ag$_2$SO$_4$ intermediate. The powder was dispersed and tested alongside g-C$_3$N$_4$, in a 2 mM FeCl$_3$ solution; at which point water splitting did not occur. Therefore it was concluded Ag$_3$PO$_4$ was not suitable for a water splitting reaction in a Fe$^{2+/3+}$ mediator.

Ag$_3$PO$_4$ was subsequently tested in system where NaI (I$^-$/IO$_3^-$) acted as the redox mediator, and g-C$_3$N$_4$ was the hydrogen production photocatalyst. A 5 mM NaI aqueous solution prepared, and tetrahedral Ag$_3$PO$_4$ immersed in the electrolyte. The pH was monitored, and increased from 8.3, to 10.5; at which point the colloid began to change colour from bright yellow, to a light grey/brown. After attempting to collect the spent Ag$_3$PO$_4$, only a tiny amount was recoverable—the remaining dissolved in the solution. The recovered powder was too little in volume for XRD, so TEM-EDX was performed in order to ascertain the powders nature.

Figure 5.2a shows a small crystalline region of AgI, comprising of two visible atomic spacings, 0.27 and 0.37 nm, corresponding to the (002) and (101) crystal diffraction planes of wurtzite AgI (P63mc, JCPDS 78-1614). The equivalent FFT (Fig. 5.1b) also shows 'diffraction' spots at the same inverse distances (see Table 5.1 for conversion).

Figure 5.3a Shows the atomic spacing of regular Ag$_3$PO$_4$ on a TEM micrograph; 0.43 nm corresponding to the (110) crystal plane. Figure 5.3b is the equivalent FFT of the TEM micrograph, which shows the 'diffraction' spots of the (110) plane. The conversion of inverse length to real space length is shown in Table 5.2

Figure 5.4a displays the full elemental spectrum of a Ag$_3$PO$_4$-AgI sample, and Fig. 5.4b illustrates the distribution of elemental Iodine across the sample. From Figs. 5.2, 5.3, and 5.4 it is apparent that the AgI layers are both in clumps and well

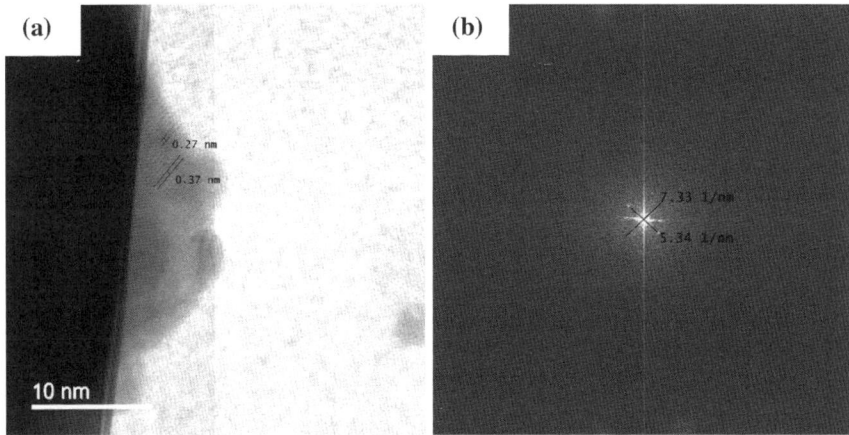

Fig. 5.2 **a** TEM micrographs of Ag_3PO_4 with AgI layer post water splitting, in g-C_3N_4–I^-/ IO_3^-–Ag_3PO_4 system. Atomic spaces indicated by parallel lines. **b** Corresponding FFT of micrograph, diameter of diffraction spots are indicated

Table 5.1 AgI FFT diameter relation to a real distance

AgI			
FFT diameter (1/nm)	FFT radius (1/nm)	Real distance (nm)	Real distance (Å)
7.32	3.66	0.27	2.73
5.33	2.67	0.38	3.75

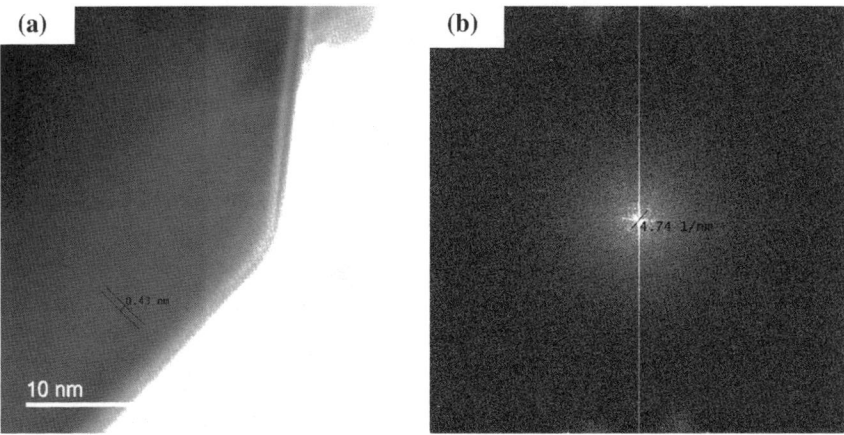

Fig. 5.3 **a** TEM micrograph of an Ag_3PO_4 crystal post water splitting, in g-C_3N_4–I^-/IO_3^-– Ag_3PO_4 system. Atomic plane distance indicated by parallel lines. **b** Corresponding FFT of micrograph, diameter of diffraction spot is indicated

Table 5.2 Ag₃PO₄ FFT diameter relation to a real distance

Ag₃PO₄

FFT diameter (nm⁻¹)	FFT radius (nm⁻¹)	Real distance (nm)	Real distance (Å)
4.700	2.350	0.426	4.255

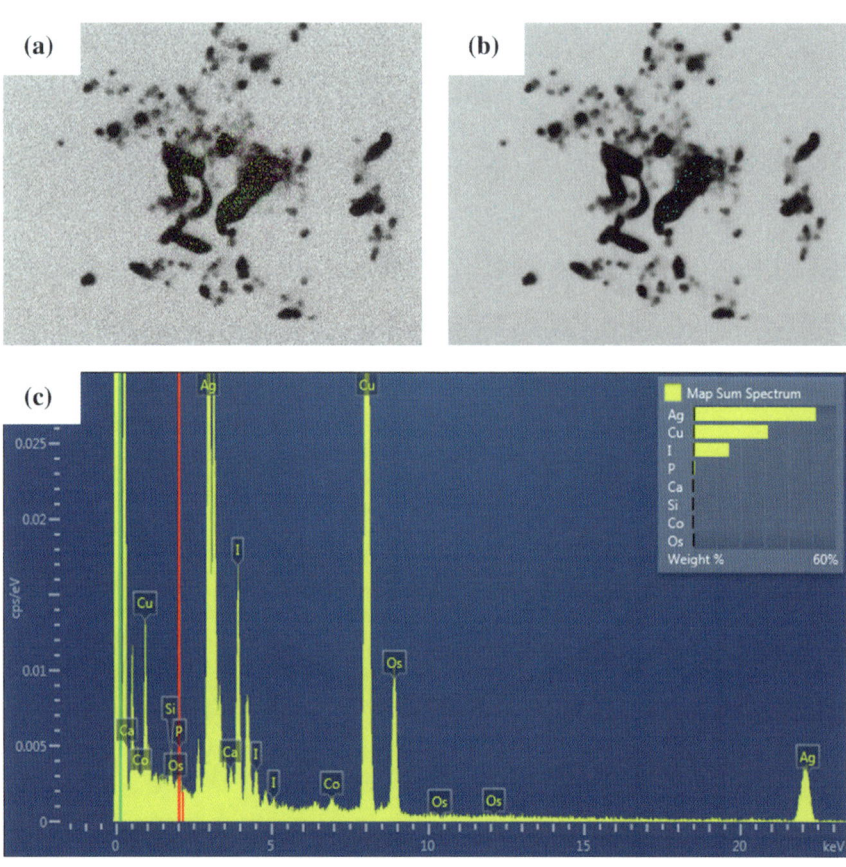

Fig. 5.4 TEM micrographs (low res mode) combined with EDX of Ag₃PO₄-AgI post water splitting in g-C₃N₄–I⁻/IO₃⁻–Ag₃PO₄ system. **a** Full spectrum mapping of TEM-EDX micrograph (Ag = *green*, O = *pink*, P = *yellow*, I = *blue*). **b** Iodine mapping of TEM-EDX micrograph. **c** EDX spectrum of corresponding TEM micrograph area (5 × 5 μm), inset denotes weight % of map sum spectrum. *Note* Cu signal is from the copper/holey carbon grid used as a support, and other heavy elements such as Ca, Si, Co, Os (all 0 wt%) are artefacts

distributed across the Ag₃PO₄ surface. Whilst not detrimental to the overall structure of Ag₃PO₄ particles, AgI layers form quickly, and then are in turn partially reduced (to change from a yellow, to light greg hue) to elemental silver, a process which stops Ag₃PO₄ from absorbing light, and halts photocatalytic activity. The ion-exchange of phosphate ions into the solution with iodine ions could also be the reason that the pH increase towards more basic values and increases the solubility of Ag_3PO_4, which could be why very little is recoverable.

Ignoring copper and other elements denoted as artefacts (0 % weight but show up in the spectrum regardless), Fig. 5.4c shows that by weight, there is a significant amount of iodine in the sample. As iodine has a large atomic weight, this value (ca. 12 %) in terms of molar amount is obviously skewed, and would be much lower (ca. 4 %).

5.3.2 Graphitic Carbon Nitride Based Z-Scheme Water Splitting Systems

Figure 5.5a shows the XRD data of synthesized g-C_3N_4, $BiVO_4$, and WO_3. Urea derived g-C_3N_4 exhibits the usual weak diffraction peaks at $13.0°$ (d = 0.681 nm) and $27.4°$ (d = 0.326 nm), corresponding to the approximate dimension of the tri-s-triazine (heptazine) unit, and the distance between graphitic layers respectively [8]. The phase and purity of the synthesised $BiVO_4$ was found to be consistent with the parent literature; pure phase monoclinic point group, space group I2/a [6]. Commercial WO_3 was found to be phase identical to photocatalytically active compounds mentioned in the literature; pure phase monoclinic point group, space group P21/n [9, 10]. Figure 5.6 shows the SEM and TEM micrographs of these semiconductors. WO_3 particles are large agglomerates (ca. 10 μm), consisting of smaller particles ca. 1 μm (Fig. 5.6a, b). $BiVO_4$ particles range from 100 nm upwards to 0.5 μm, in dendritic agglomerates (Fig. 5.6c). Graphitic carbon nitride is a porous sheet-like compound, with has no fixed particle size (Fig. 5.6d, e). UV-Vis spectra in Fig. 5.5b shows all materials have visible light absorption, increasing in the order: g-C_3N_4, WO_3, and $BiVO_4$; in agreement with the previous reports and signifying the different systems have suitable band gaps for half-reactions [11, 12].

Initially, Pt-g-C_3N_4 was tested for water splitting in the absence of redox mediators as shown in Table 5.3 (run 1), to which there was no water splitting activity. Pt-loaded g-C_3N_4 (via photodeposition) is shown to be active for both oxidation of NaI (I^-, run 2) and $FeCl_2$ (Fe^{2+}, run 3) and hydrogen is produced more readily and the relevant chemical reactions were mentioned in Sect. 1.3.3. No oxygen evolution is observed. Then, Pt-g-C_3N_4 was tested for water splitting using only redox mediators which can be oxidized. When coupled with a redox mediator, and in combination with either Pt-loaded WO_3 (via impregnation, run 4) or $BiVO_4$, pure water splitting is observed.

Water splitting over Pt-g-$C_3N_4/BiVO_4$ in an aqueous $Fe^{2+/3+}$ solution was tested with different starting oxidation states of the redox precursors; $FeCl_2$ (run 6) and $FeCl_3$ (run 7). Starting with Fe^{2+} species, pure water splitting under full arc commences instantly, and after an initial 1 h of equilibration, continues steadily. However, if the starting species is Fe^{3+}, the overall water splitting reaction does

Fig. 5.5 **a** Powder XRD
pattern and **b** UV-Vis
absorbance spectra of
g-C$_3$N$_4$, WO$_3$ and BiVO$_4$
compounds prior to water
splitting reactions

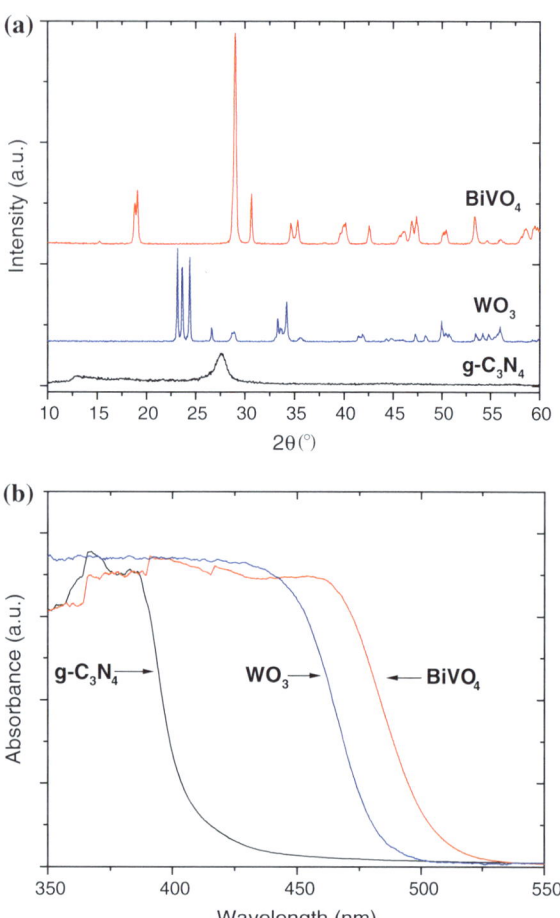

not occur instantly apart from only O$_2$ evolution. After 15 h, and re-purging for
2 h, water splitting occurs due to sufficient Fe^{2+} existed in the system produced
by the previous 15 h photoreduction process, indicating this side of the reaction
proceeds much slower (reduction of Fe^{3+} by CB electrons in both g-C$_3$N$_4$ and
BiVO$_4$). This also indicates that Fe^{3+} can be regenerated in the present system. As
shown in Fig. 5.7 experimental optimization procedures were performed, which
involved changing pH, photocatalyst amount, and FeCl$_2$ concentration.

It was found that by varying the concentration of FeCl$_2$ in the aqueous solution,
water splitting rates can be optimized. At 0.5 mM FeCl$_2$, the water splitting rates
are 3.5 and 7.1 μmol h^{-1} g^{-1} for oxygen and hydrogen respectively. Increasing
the concentration to 2 mM shows that the initial water splitting rate double to
7 and 13.5 μmol h^{-1} g^{-1} for oxygen and hydrogen respectively. Increasing the
concentration to 4 mM FeCl$_2$ then decreases water splitting rates to 4.5 and
9 μmol h^{-1} g^{-1} for oxygen and hydrogen respectively. Although no definitive

Fig. 5.6 SEM and TEM micrographs of different photocatalyst. **a** & **b** Commercial WO_3 crystals. **c** Synthesized $BiVO_4$ crystals. **d** & **e** Synthesized g-C_3N_4 sheets

theory has been published to explain this phenomenon [13], all further optimization procedures were conducted using 2 mM $FeCl_2$ as it the most efficient concentration for overall water splitting.

The mass ratio between photocatalysts is also crucial to the activity, reaching a maximum at a ratio of 2:1 (g-C_3N_4:$BiVO_4$), however, the hydrogen and oxygen evolution rates deviate from stoichiometry, presumably to the larger surface area of g-C_3N_4 compared to $BiVO_4$ (43.8 m^2 g^{-1} vs. 3.9 m^2 g^{-1} respectively) and better dispersion in a solution. At 1:1, there is stoichiometry between oxygen and hydrogen, and the rates peak at 7 and 15 μmol h^{-1} g^{-1} respectively.

Table 5.3 Overall water splitting under full arc irradiation (300 W Xe lamp) using different redox mediated systems

Run	H$_2$ photocatalyst (all 3 wt% Pt)	O$_2$ photocatalyst	Weight ratio (g:g)	pH	Redox mediator	Initial gas evolution rate (μmol h^{-1} g^{-1}) H$_2$	O$_2$
1	g-C$_3$N$_4$	–	–	7	–	0	–
2	g-C$_3$N$_4$	–	–	8.3	NaI (5 mM)	5	–
3	g-C$_3$N$_4$	–	–	3	FeCl$_2$ (2 mM)	4	–
4	–	BiVO$_4$	–	3	FeCl$_3$ (2 mM)	–	90
5	–	WO$_3$ (0.5 wt% Pt)	–	7	NaIO$_3$ (5 mM)	–	35
6	g-C$_3$N$_4$	BiVO$_4$	1:1	3	FeCl$_2$ (2 mM)	15	8
7	g-C$_3$N$_4$	BiVO$_4$	1:1	3	FeCl$_3$ (2 mM)	–	3
8	g-C$_3$N$_4$	WO$_3$ (0.5 wt% Pt)	1:1	8.3	NaI (5 mM)	74	37

Since the pH of Fe$^{2+/3+}$ systems have previously reported to be sensitive, due to iron ion hydrolysis, we can see an expected optimum at pH 2.4–3 [14]. Post water splitting XRD data (Fig. 5.8), however, shows that BiOCl is formed as a sub-phase during the reaction at pH 2.4, whilst at pH 3, BiVO$_4$ did not suffer any degradation. Therefore further testing of the g-C$_3$N$_4$ (3 wt% Pt)—BiVO$_4$ system was conducted at pH 3.

As shown in Fig. 5.9a, b, under full arc and visible light, both H$_2$ and O$_2$ evolve linearly after an initial 1 h period of equilibration, for the next 7 h with a ratio of ca. 2:1 (H$_2$ and O$_2$: 15 and 8 μmol h^{-1} g^{-1}). Under visible irradiation (a cut-off filter of 395 nm was used), an average of 6 and 3 μmol h^{-1} g^{-1} of H$_2$ and O$_2$ has been observed under optimized conditions. The water splitting rate under full arc irradiation is approximately 2.5 times larger than that of the system exposed to visible light ($\lambda > 395$ nm). However, when considering that the visible light wavelength range of the Z-scheme system is limited by the semiconductor with the largest band gap (i.e. smallest absorption window), g-C$_3$N$_4$, the wavelength range is comparatively small; $395 < \lambda < 415$ nm of the visible light system, and $300 < \lambda < 415$ nm for the full arc system. Therefore, the visible light system is actually reasonably efficient considering the wavelength range is over 5 times smaller, which affects both energy per photon, and total photon flux.

Pt-loaded g-C$_3$N$_4$ was also tested for water splitting when using NaI as a redox mediator. Pt-WO$_3$ (0.5 % as stated in many reports) [13] was used as a relatively robust photocatalyst (pH dependent) for the reduction of IO$_3^-$ species produced as a result of I$^-$ oxidation by Pt-g-C$_3$N$_4$ [15]. Again, the weight ratio of photo-catalysts, pH, and mediator concentration were carefully tailored to increase gas evolution rates from water splitting (Fig. 5.10). Whilst Fe$^{2+/3+}$ redox systems

Fig. 5.7 Optimisation of
water splitting rates using a
g-C$_3$N$_4$–BiVO$_4$ Z-scheme
system with FeCl$_2$ as a redox
mediator. **a** pH modification,
b differing concentration of
FeCl$_2$, **c** alternating the ratio
of g-C$_3$N$_4$ to BiVO$_4$

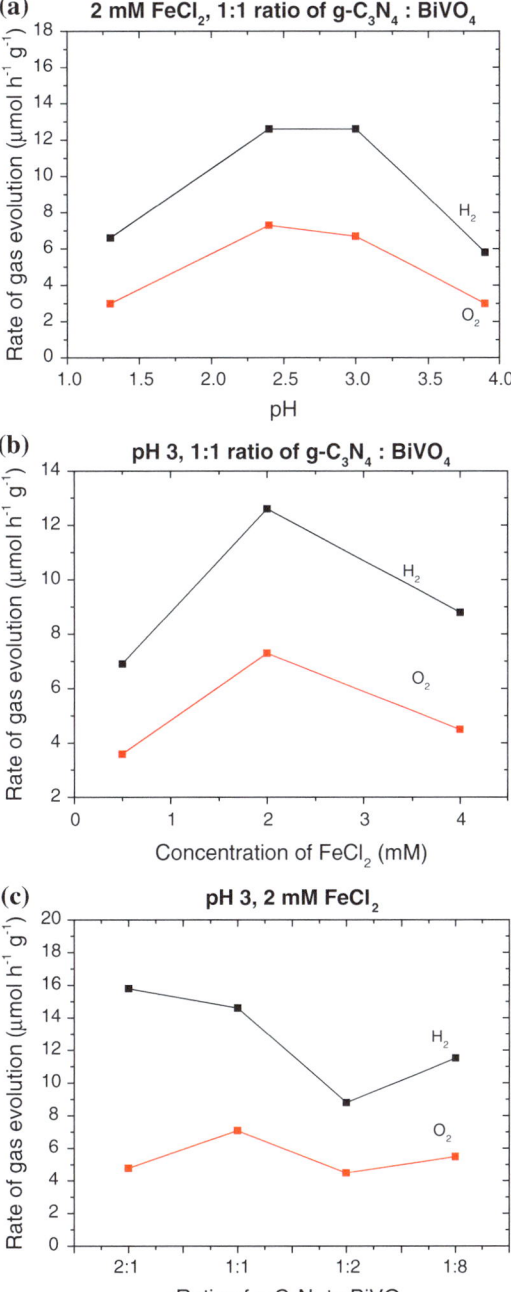

Fig. 5.8 Post water splitting XRD data of g-C$_3$N$_4$ //BiVO$_4$ system at different pH levels. Tetragonal phase BiOCl, JCPDS 00-006-0249, monoclinic phase BiVO$_4$, JCPDS 00-044-0081

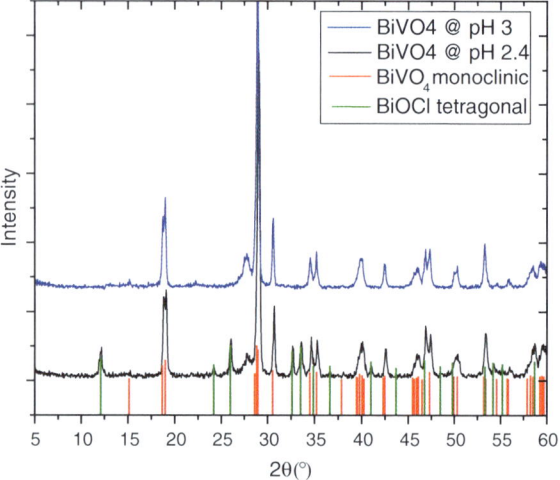

Fig. 5.9 Stoichiometric water splitting (300 W Xe lamp source): **a** g-C$_3$N$_4$ (3 wt% Pt)–FeCl$_2$–BiVO$_4$ under full arc irradiation, pH 3, 1:1 photocatalyst weight ratio, 2 mM FeCl$_2$. **b** g-C$_3$N$_4$ (3 wt% Pt)–FeCl$_2$–BiVO$_4$ under visible light irradiation (λ > 395 nm), pH 3, 1:1 photocatalyst weight ratio, 2 mM FeCl$_2$

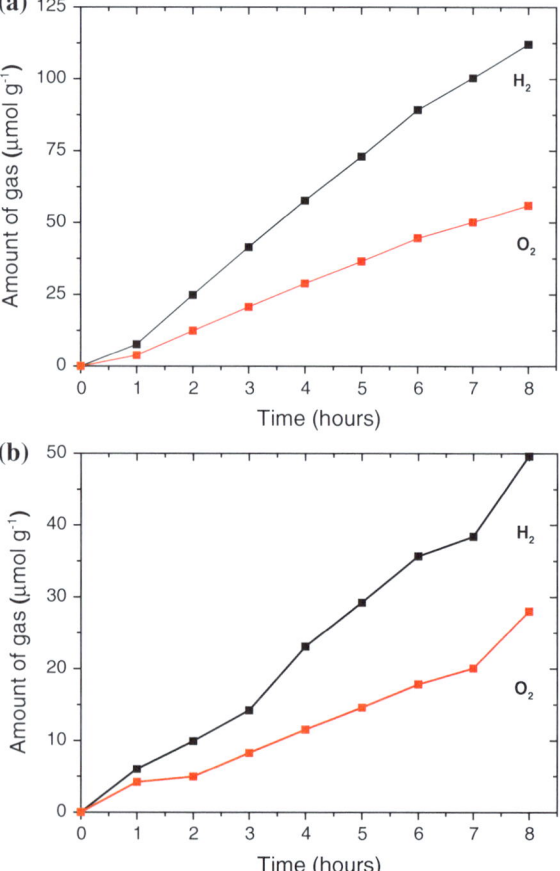

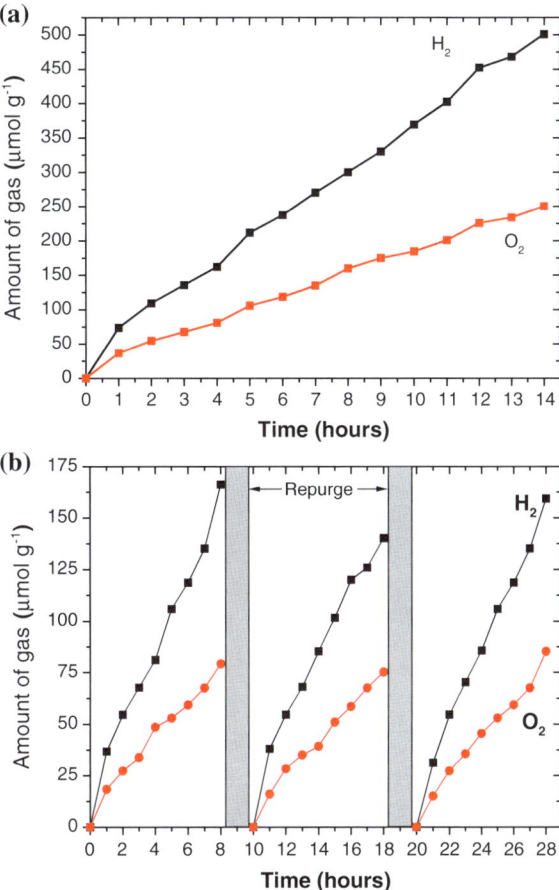

Fig. 5.10 Water splitting using **a** g-C$_3$N$_4$ (3 wt% Pt)–NaI–WO$_3$ (0.5 wt% Pt) under full arc irradiation, pH 8.3, 1:1 photocatalyst weight ratio, 5 mM NaI and, **b** g-C$_3$N$_4$ (3 wt% Pt)–NaI–WO$_3$ (0.5 wt% Pt) under visible light irradiation ($\lambda > 395$ nm), pH 8.3, 1:1 photocatalyst weight ratio, 5 mM NaI

are required to operate in acidic conditions, NaI/IO$_3$$^-$ redox systems have been reported to be stable in a pH range of 7–10, with pH 9 being optimal for titania based systems [11]. However, WO$_3$ is unstable in alkaline media around pH 9, and forms an oxyanionic tungstate compound or the more likely hydrogentungstate ion [16] (see Eq. 5.1).

$$WO_3 + OH^- \rightarrow [HWO_4]^- \tag{5.1}$$

It is therefore not surprising that the Pt-g-C$_3$N$_4$–Pt-WO$_3$ system is most efficient at pH 8.3 when using a NaI/IO$_3$$^-$ redox couple (Table 5.3, run 8). Pt-g-C$_3$N$_4$–Pt-WO$_3$ was not tested in the presence of Fe$^{2+/3+}$ redox simply because WO$_3$ is unstable in acidic conditions (pH $\leq$ 3) [16], and undergoes protonation according to:

$$WO_3 + H^+ \rightarrow H_yWO_x \tag{5.2}$$

The undesirable and ineffective I_3^- anion (Eq. 5.2.), which has shown previously to be detrimental to I^-/IO_3^- based Z-scheme systems, ceases to be produced in alkaline media [11, 17].

The build-up of the I_3^- ion in acidic media not only negates electron donation from the photocatalyst, but the ion also absorbs light (up to 350 nm), hindering photon absorption by the photocatalyst slightly.

$$3I^- + 2h^+ \rightarrow I_3^- \qquad (5.3)$$

Alternatively, in neutral and alkaline condition, I^- is oxidised to IO_3^-

$$I^- + 3H_2O + 6h^+ \rightarrow IO_3^- + 6H^+ \ (neutral\ pH) \qquad (5.4)$$

$$I^- + 6OH^- + 6h^+ \rightarrow IO_3^- + 3H_2O \ (basic\ pH) \qquad (5.5)$$

Production of I_3^- ions is also hypothesised to be suppressed by the following reaction whereby the photocatalyst is essentially inert [11]:

$$3I_3^- + 6OH^- \rightarrow 8I^- + IO_3^- + 3H_2O \qquad (5.6)$$

Therefore the pH of this system is particularly important for successful reaction kinetics; pH 8.3 is essentially a point where WO_3 is stable, and the production of I_3^- ions are suppressed, so that water splitting can occur relatively efficiently.

The weight ratio of the two photocatalysts was varied and it was found that similar to the above g-C_3N_4–$BiVO_4$ system, this ratio influences the stoichiometry of H_2 and O_2 production. The optimum weight ratio is 1:1. At 1:1 weight ratio, stable water splitting rates are observed. Averaging over 14 h, 36 and 18 μmol h^{-1} g^{-1} of H_2 and O_2 are produced, and linearly increase under full arc, shown in Fig. 5.10a. Under visible irradiation ($\lambda > 395$ nm), an average of 21.2 and 11.0 μmol h^{-1} g^{-1} of H_2 and O_2 evolve and the activity is very stable, as indicated by three cycles in Fig. 5.10b. If the ratio is raised beyond this, water splitting rates decrease, presumably because of an imbalance in light absorption between respective photocatalysts, likely due to a difference in refractive index between the two materials which thus affects the number of photons scattered/reflected. For example, when using 0.3 g of Pt-g-C_3N_4, the hydrogen production rate is initially higher, yet eventually drops off; this is attributed to Pt-WO_3 not absorbing the equivalent photon flux as Pt-g-C_3N_4 (Fig. 5.11). Figure 5.12 illustrates the unchanged crystal structure of the system post 24 h water splitting; demonstrating the two organic semiconductor-based systems are extremely stable for pure water splitting.

By far the most efficient system in this study was g-C_3N_4–NaI–WO_3, and the solar-to-hydrogen energy conversion efficiency was calculated (STH%, Eq. 1.10). Using a 150 W Xe lamp equipped with an AM 1.5 global filter, 100 mW cm^{-2}, the current solar energy conversion efficiency is 0.1 %.

Despite participating in the most active system for overall water splitting using g-C_3N_4, it was postulated that the non-optimal morphology and surface

Fig. 5.11 Optimisation of
g-C$_3$N$_4$–WO$_3$ Z-scheme
system with NaI as a redox
mediator. **a** pH modification,
b differing concentration of
NaI, **c** alternating the ratio of
g-C$_3$N$_4$ to WO$_3$

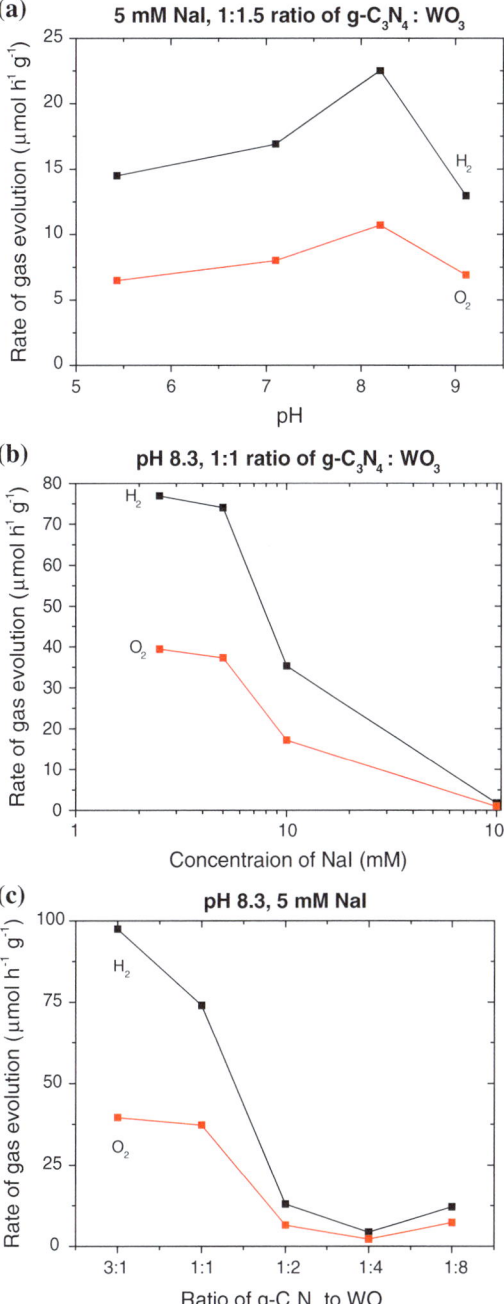

Fig. 5.12 Post XRD
diffraction patterns from two
different g-C₃N₄ based water
splitting systems

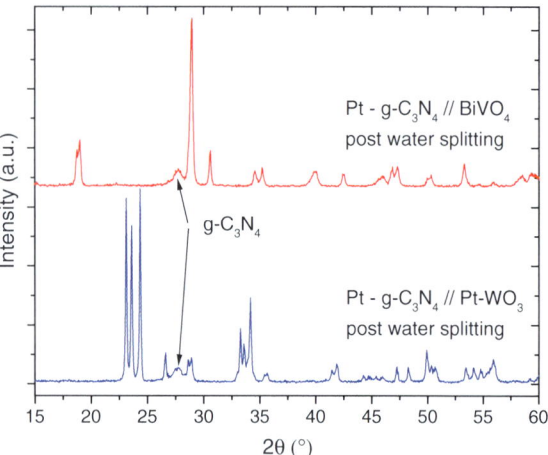

Fig. 5.13 BET isotherms
of commercial WO₃
(SSA = 1.15 m²g⁻¹), BiVO₄
(SSA = 3.91 m²g⁻¹), and
WO₃ platelets (2.45 m²g⁻¹)

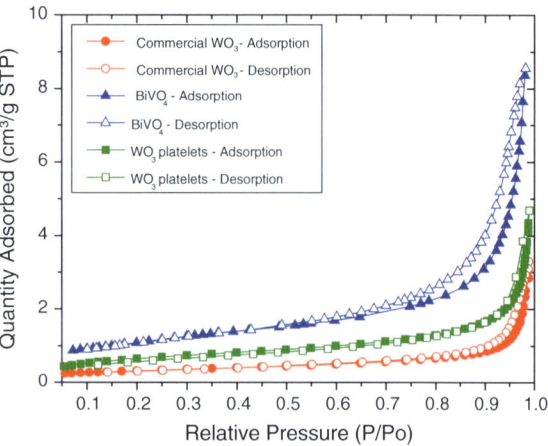

area of WO₃ (Fig. 5.6a, b, and 5.13), could be optimised in order to enhance the efficiency. WO₃ micro/nanoplates with preferential {001} facets were synthesised, phase identical to that of commercial monoclinic WO₃ (JCPDS 01-083-0951). As seen in Fig. 5.14, the XRD pattern of the two compounds matches in terms of peak position, indicating the same monoclinic phase. However, upon closer inspection, we see that the intensity ratio of the (002), (020) and (200) peaks at 23.15, 23.6 and 24.4° 2θ respectively, vary between samples in comparison to the standard commercial WO₃ pattern. The (002) Bragg peak is much more pronounced in comparison to both (020) and (200) peaks on the as-synthesised platelet WO₃. As shown in Table 5.4, the intensity ratio of (002)/(020) is 2.50 for platelet WO₃, whereas it is only 1.26 for commercial, non-faceted WO₃. Similarly the ratio

Fig. 5.14 Comparison of XRD diffraction patterns of WO_3 commercial crystals and as synthesised WO_3 platelets

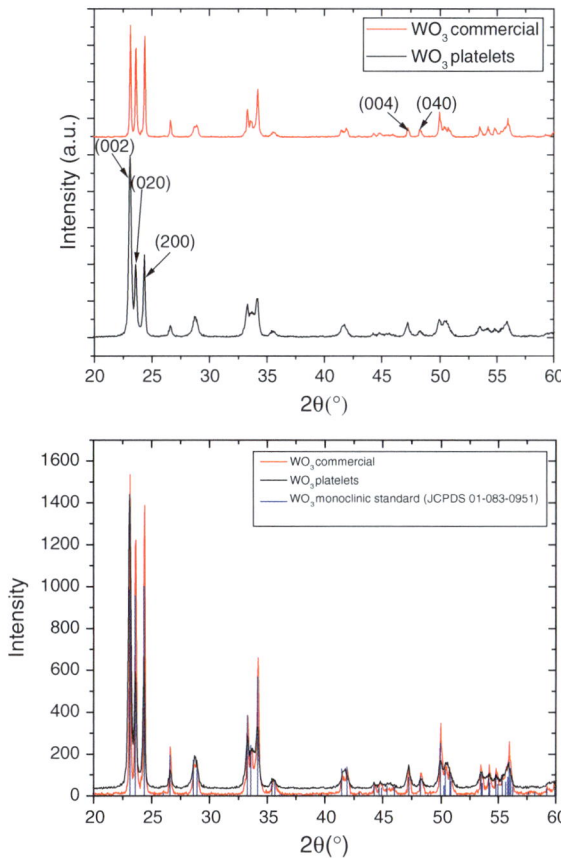

Table 5.4   Intensity ratios of WO_3 PXRD Bragg peaks

	Peak intensity (2θ, reflection plane)				
Sample	21.3°, (002)	23.6°, (020)	24.35°, (200)	47.3°, (004)	48.3°, (040)
Standard	3839	3779	3790	417	426
WO_3 commercial	1535	1223	1387	133	110
WO_3 platelets	2489	996	1129	210	91
	(002)/(002)	(002)/(020)	(002)/(200)	(004)/(004)	(004)/(040)
Ratio standard	1.00	1.02	1.01	1.00	0.98
Ratio WO_3 commercial	1.00	1.26	1.11	1.00	1.21
Ratio WO_3 platelets	1.00	2.50	2.20	1.00	2.31

Fig. 5.15 Low (**a**) and high (**b**) SEM micrographs of platelet WO_3

of (002)/(200) for platelet WO_3 is 2.20, and only 1.11 for commercial WO_3. This preferential crystal orientation of the (002) plane is evident from SEM micrographs, which demonstrate the plate-like nature (Fig. 5.15).

0.5 wt% platinum was deposited as a cocatalyst on the surface, and the platelet WO_3 was tested in water splitting system in combination with Pt-g-C_3N_4 and 5 mM NaI acting as a redox mediator. No hydrogen or oxygen was evolved during a 12 h period; water splitting did not occur. This is surprising, however, a hypothesis could be made when considering evidence from absorption spectra and Tauc plots of the two variants of WO_3 (Fig. 5.16). Due to the platelet WO_3 particles increased absorption, presumably from the exposing terminated/faceted surface, and smaller indirect band gap [18] (2.48 eV in comparison to 2.62 eV, Fig. 5.16b, d), the conduction band edge could be pushed to more positive potentials by ~0.2 eV. When considering the energetic requirements needed to reduce IO_3^- to I^- (redox potential of $I^-/IO_3^- = +0.67$ eV [19], at pH 7) and subsequently liberate oxygen in a water splitting system, it is conceivable that this conduction band edge shift (from CBE of commercial WO_3 +0.6 eV [20] to that of platelet sample +0.8 eV at pH 7) has negated the ability of WO_3 to reduce the redox mediator, IO_3, and thus electron hole recombination dominates within the photocatalyst, disabling the production of oxygen from water. The Tauc plot in Fig. 5.16c also show that the WO_3 platelets might possess a direct transition of 2.35 eV, which might also explain why there is no activity—as the band gap has many sub-band gap states which might act as electron traps and enforce charge recombination.

The change in band gap between the two WO_3 samples directly indicates how sensitive the redox mediator-photocatalyst coupling is; a small change in band gap can have a dramatic effect on the effectiveness of the overall water splitting system.

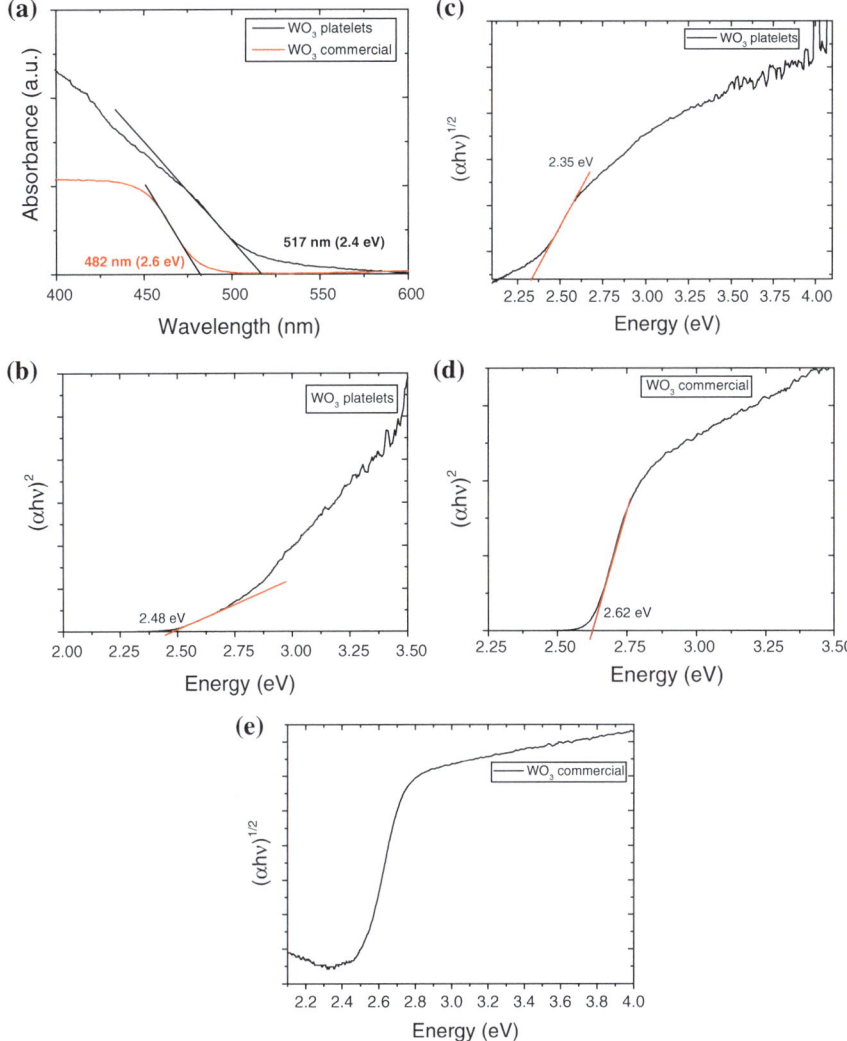

Fig. 5.16 a UV-Vis absorption spectra of commercial WO_3, and platelet WO_3 particles, Tauc plots of; WO_3 platelets (**b** & **c**), and WO_3 commercial particles (**c** & **d**)

5.4 Conclusions

To summarise, two novel g-C_3N_4 based systems for pure water splitting under both UV and visible light irradiation have been demonstrated. The most efficient system is composed of g-C_3N_4 and WO_3 using I^-/IO_3^- as a soluble redox mediator, which yields 36 and 18 μmol h^{-1} g^{-1} of hydrogen and oxygen respectively. The stability of the system is evidenced in the linear gas evolution over 28 h. This leads to a STH% of ca. 0.1 %. Both $BiVO_4$ ($Fe^{2+/3+}$) and WO_3 (I/IO_3^-) based systems were shown to be heavily dependent on pH, weight ratio between photocatalysts, and redox mediator concentration. The energetic sensitivity of the WO_3 (I/IO_3^-) system was demonstrated by testing WO_3 platelets, which have a smaller band gap than their commercial counterparts, and this was shown to have a dramatic negative effect upon the system—no water splitting was observed. However, the findings pave the way for more systems to be created that are based on urea-derived g-C_3N_4, provided the band positioning of the semiconductor is carefully controlled.

Carbon nitride synthesized using other precursors such as dicyandiamide or thiourea was also tested for water splitting using both NaI–WO_3 and $FeCl_2$–$BiVO_4$ systems, yet do not give detectable activity, conceivably due to the low efficiency of these carbon nitrides in comparison to urea-derived g-C_3N_4. Low efficiency as a result of a smaller driving force/band gap could lead to an imbalance in redox mediation species, or become comparative to competitive side reactions (I^- oxidation to I_3^- instead of IO_3^-) which would eventually halt the water splitting reaction. It is well documented that redox mediator ions such as I/IO_3^- and $Fe^{2+/3+}$ are particularly sensitive when adsorbing to surfaces, as explored by Abe et al., and Ohno et al. [11, 21]. The studies show that the IO_3^- ion preferentially adsorbs on to a rutile surface rather than anatase, and is effectively reduced, meanwhile water is effectively oxidised by rutile TiO_2. Considering anatase and rutile have different isoelectric points(anatase ca. 5, rutile <3) [22], and therefore surface charge, it is possible the reason behind the selective absorption is due to zeta potential at the surface. The same premise could apply with g-C_3N_4, since the degree of protonation or surface acidity varies from sample to sample. In which case, the conjecture would be that I^- ions selectively bind to urea based g-C_3N_4 due to the relatively small proton concentration, and it is more difficult to bind to DCDA or thiourea based carbon nitride as the potential is larger.

It was found that Ag_3PO_4 was not able to participate in a Z-scheme system whereby $FeCl_3$ or NaI are the redox precursors, due to interaction with the mediator, and unsuitable pH ranges.

References

1. Maeda, K. et al. (2006). Photocatalyst releasing hydrogen from water. *Nature 440*, 295–295.
2. Ferreira, K. N., Iverson, T. M., Maghlaoui, K., Barber, J., & Iwata, S. (2004). Architecture of the photosynthetic oxygen-evolving center. *Science, 303*, 1831–1838.
3. Bowker, M. (2011). Sustainable hydrogen production by the application of ambient temperature photocatalysis. *Green Chemistry, 13*, 2235–2246.
4. Bard, A. J. (1979). Photoelectrochemistry and heterogeneous photo-catalysis at semiconductors. *Journal of Photochemistry, 10*, 59–75.
5. Kato, H., Hori, M., Konta, R., Shimodaira, Y., & Kudo, A. (2004). Construction of Z-scheme Type heterogeneous photocatalysis systems for water splitting into H_2 and O_2 under visible light irradiation. *Chemistry Letters, 33*, 1348–1349.

6. Kudo, A., Omori, K., & Kato, H. (1999). A novel aqueous process for preparation of crystal form-controlled and highly crystalline BiVO$_4$ powder from layered vanadates at room temperature and its photocatalytic and photophysical properties. *Journal of the American Chemical Society, 121*, 11459–11467.

7. Patnaik, P. (2003). *Handbook of Inorganic Chemicals*. New York: McGraw-Hill.

8. Wang, X., et al. (2008). A metal-free polymeric photocatalyst for hydrogen production from water under visible light. *Nature Materials, 8*, 76–80.

9. Darwent, J. R., & Mills, A. (1982). Photo-oxidation of water sensitized by WO$_3$ powder. *Journal of the Chemical Society, Faraday Transactions 2: Molecular and Chemical Physics, 78*, 359–367.

10. Zheng, H., et al. (2011). Nanostructured tungsten oxide—properties, synthesis, and applications. *Advanced Functional Materials, 21*, 2175–2196.

11. Abe, R., Sayama, K., & Sugihara, H. (2005). Development of new photocatalytic water splitting into H$_2$ and O$_2$ using two different semiconductor photocatalysts and a Shuttle redox mediator IO$_3^-$/I$^-$. *The Journal of Physical Chemistry B, 109*, 16052–16061.

12. Kato, H., Sasaki, Y., Iwase, A., & Kudo, A. (2007). Role of iron ion electron mediator on photocatalytic overall water splitting under visible light irradiation using Z-scheme systems. *Bulletin of the Chemical Society of Japan, 80*, 2457–2464.

13. Maeda, K. (2013). Z-Scheme water splitting using two different semiconductor photocatalysts. *ACS Catalysis, 3*, 1486–1503.

14. Kato, H., Sasaki, Y., Iwase, A., & Kudo, A. (2007). Role of iron ion electron mediator on photocatalytic overall water splitting under visible light irradiation using Z-scheme systems. *Bulletin of the Chemical Society of Japan, 80*, 2457–2464.

15. Maeda, K., Higashi, M., Lu, D., Abe, R., & Domen, K. (2010). Efficient nonsacrificial water splitting through two-step photoexcitation by visible light using a modified oxynitride as a hydrogen evolution photocatalyst. *Journal of the American Chemical Society, 132*, 5858–5868.

16. Enesca, A., Andronic, L., Duta, A., & Manolache, S. (2007). Optical properties and chemical stability of WO$_3$ and TiO$_2$ thin films photocatalysts. *Romanian Journal of Information Science and Technology, 10*, 269–277.

17. Ohno, T., Saito, S., Fujihara, K., & Matsumura, M. (1996). Photocatalyzed production of hydrogen and iodine from aqueous solutions of iodide using platinum-loaded TiO$_2$ powder. *Bulletin of the Chemical Society of Japan, 69*, 3059–3064.

18. González-Borrero, P. P. et al. (2010). Optical band-gap determination of nanostructured WO$_3$ film. *Applied Physics Letters 96*. doi:http://dx.doi.org/10.1063/1061.3313945.

19. Abe, R., Sayama, K., Domen, K., & Arakawa, H. (2001). A new type of water splitting system composed of two different TiO$_2$ photocatalysts (anatase, rutile) and a IO$_3^-$/I$^-$ shuttle redox mediator. *Chemical Physics Letters, 344*, 339–344.

20. Cabrera, R. Q., et al. (2012). Photocatalytic activity of needle-like TiO2/WO3−x thin films prepared by chemical vapour deposition. *Journal of Photochemistry and Photobiology A: Chemistry, 239*, 60–64.

21. Ohno, T., Sarukawa, K., & Matsumura, M. (2002). Crystal faces of rutile and anatase TiO$_2$ particles and their roles in photocatalytic reactions. *New Journal of Chemistry, 26*, 1167–1170.

22. Mandzy, N., Grulke, E., & Druffel, T. (2005). Breakage of TiO$_2$ agglomerates in electrostatically stabilized aqueous dispersions. *Powder Technology, 160*, 121–126.

Chapter 6
Overall Conclusions and Future Work

6.1 Overall Conclusions

A wide variety of literature was studied in Chap. 1 in order to identify the benchmark materials developed and more importantly the barrier which limits the solar to H_2 evolution efficiency. Novel strategies were then attempted in order to tackle the difficulties experienced in the field.

The first publication on Ag_3PO_4 by Yi et al. displayed great promise, demonstrating a relatively good quantum yield for a novel photocatalyst, but the efficiency was not high enough, in particular at long wavelength (e.g. 500 nm) for the photooxidation of water [1]. It was unclear as to what would further augment or diminish the activity, and thus was selected to be studied for water photooxidation. In parallel, graphitic carbon nitride was also thought to be promising as a visible light hydrogen production photocatalyst. Despite the relatively low quantum yield reported in the first instance, the compound is very stable chemically, and also is incredibly cheap to synthesise. Considering these arguments, it was decided that the understanding and modification of $g-C_3N_4$ would be pursued. Z-scheme photocatalytic systems have been reported, yet for only a relatively small number of photocatalysts. It was later postulated that either $g-C_3N_4$ or Ag_3PO_4 could potentially be integrated into a Z-scheme system, based on their band structure and good photocatalytic activity.

Firstly the development of a reliable water splitting test station was described; a combination of both a custom designed, gas tight, borosilicate reactor, and the integration of this with a gas chromatograph. The gas chromatography unit was calibrated to accurately determine unknown quantities of hydrogen and oxygen, down to the nanomole scale, and with a small sampling error of 0.52 and 0.97 % for hydrogen and oxygen respectively.

Then the focus of the thesis is directed towards the development of efficient water photooxidation using a Ag_3PO_4 photocatalyst. A method was found whereby to roughly spherical silver phosphate exhibited a photooxidative ability analogous to that published by Yi et al., using ethanol, $AgNO_3$, and Na_2HPO_4. Upon experimentation with phosphate precursors, it was found that H_3PO_4 and

© Springer International Publishing Switzerland 2015

D.J. Martin, *Investigation into High Efficiency Visible Light Photocatalysts for Water Reduction and Oxidation*, Springer Theses,

DOI 10.1007/978-3-319-18488-3_6

the relative concentration could slow the growth of Ag_3PO_4 crystals, and retain low index facets, resulting in the production of novel {111} terminated tetrahedral Ag_3PO_4 crystals. By changing the concentration of H_3PO_4, and thus limiting the kinetics of growth, it is possible to control the growth of Ag_3PO_4 crystals from tetrapods to tetrahedrons. Ag_3PO_4 tetrahedrons demonstrate considerably higher oxygen evolution activity in comparison to roughly spherical particles, and other low index facets such as {100} and {100}. DFT calculations were used to model the low index surfaces surface energy, and the hole mass of Ag_3PO_4 in different directions. It is concluded that a combination of highest surface energy and lowest hole mass on Ag_3PO_4 {111} surfaces results in an extremely high oxygen evolution from water, with an internal quantum yield of nearly unity at 400 nm and 80 % at 500 nm.

The thesis then investigates $g\text{-}C_3N_4$ as a hydrogen producing photocatalyst. Graphitic carbon nitride was synthesised via thermal decomposition, using 4 different precursors, at various calcination temperatures, calcination ramp rates, and varying cocatalyst element and weighting. A new synthesis method for the production of urea derived graphitic carbon nitride was developed. The most active sample was synthesised at 600 °C, at 5 °C/min, and by using 3 wt% platinum as a cocatalyst. This sample evolved hydrogen from water in the presence of TEOA at 3327.5 μmol h^{-1} g^{-1} under visible light (l > 395 nm), and ca. 20,000 μmol g^{-1} under full arc irradiation. The internal quantum yield in the visible region is 26.5 %, nearly ten times larger than previous literature examples, or the sample synthesised using MCA/DMSO. In order to determine the reason for the high reactivity, in comparison to other examples, a series of characterisation methods were employed. TGA-DSC-MS was used to monitor the formation mechanism of $g\text{-}C_3N_4$ derived from both urea and thiourea which was found to be drastically different during the primordial stages of synthesis in comparison to cyanamide [2] and thiourea. In the case of urea derived $g\text{-}C_3N_4$, it is shown that the high surface area and porosity is created by the continuous loss of CO_2, a result not seen before. Formaldehyde (CH_2O) is also continually produced throughout the decomposition of urea to $g\text{-}C_3N_4$, a process which is proposed to reduce the percentage of hydrogen/protons in carbon nitride. Only trace amounts of CS_2 and CH_4N_2 remain at the point of polymerisation (500 °C) in the thermal condensation of thiourea derived $g\text{-}C_3N_4$, therefore at the point of polymerisation, the compound is not losing extra hydrogen/protons from fragmented products.

Using the N1s core level data obtained from XPS, it is shown that across all synthesised samples, there is an inversely proportional relationship between hydrogen/proton content in the photocatalyst and hydrogen production rate achieved by the photocatalyst. A larger hydrogen/proton content leads to a lower hydrogen evolution rate. This is also confirmed by Elemental Analysis, which although is not specific as to where the hydrogen is located, the trend is the same. It has been shown in previous studies that a less polymeric carbon nitride intrinsically has more hydrogen/protons due to breakages. The trend in XRD data shows that a greater level of condensation, and thus a less polymerised $g\text{-}C_3N_4$, leads to a less active photocatalyst.

DFT calculations were used to show that increasing the hydrogen/proton content causes a positive shift in the conduction band edge (with respect to NHE), therefore reducing the driving force for reduction reactions. Furthermore, TDDFT calculations demonstrate that excess protonation causes excited photoelectrons to become localised at non-active redox sites, which is further detrimental to the photocatalytic ability since these charge carriers cannot participate in water splitting reactions.

Due to the progress made in the previous two chapters, the feasibility of the integration of either Ag_3PO_4 or g-C_3N_4 into a Z-Scheme for water splitting was then assessed. Ag_3PO_4 was found to not function as an oxygen evolving photocatalyst in Z-scheme water splitting systems due to the incompatibility between photocatalyst and synthetic conditions. Ag_3PO_4 dissolves into the solution at the pH range (≤ 3.5) required for $Fe^{2+/3+}$ to be an effective redox mediator, and also reacts with H_2SO_4 at pH 4 to form an Ag_2SO_4–Ag_3PO_4 mixed phase compound (evidenced by XRD) which shows no activity for water splitting. When NaI is used, Ag_3PO_4 reacts with the iodide anion to form AgI as seen in TEM and EDX, visibly changing the colour of the solution and also demonstrating no activity for overall water splitting.

Urea derived g-C_3N_4 is shown to actively participate in novel Z-Scheme water splitting systems as a hydrogen evolving photocatalyst. Due to the robustness and favourable band position, g-C_3N_4 can be coupled with either I^-/IO_3^- or $Fe^{2+/3+}$ redox mediators, at any pH, and with Pt-loaded WO_3 or $BiVO_4$ respectively. The highest water splitting rates were achieved using a g-C_3N_4 NaI–WO_3 system, peaking at 36 and 18 μmol h^{-1} g^{-1} of hydrogen and oxygen respectively. The largest STH% was recorded at 0.1 % under AM 1.5 illumination. This is far from the commercial target of 10 % for overall water splitting systems, yet for a novel system, can be built on and improved. Carbon nitride fabricated by thermal decomposition of dicyandiamide or thiourea do not give detectable activity for any Z-scheme water splitting reaction, possibly due to the low efficiency of these carbon nitrides in comparison to urea-derived g-C_3N_4 [3]. Low efficiency as a result of a smaller driving force/higher degree of protonation could lead to an imbalance in redox mediation species, or become comparative to competitive side reactions (I^- oxidation to I_3^- instead of IO_3^-) which would eventually halt the water splitting reaction. It may also be possible that I^- ions could selectively bind to urea based g-C_3N_4 due to the relatively small proton concentration, and it is more difficult to bind to DCDA or thiourea based carbon nitride as the surface zeta potential is larger.

6.2 Future Work

The thesis achieved the main goals set out, by adding novel scientific understanding to the nature of the activity demonstrated by some of the newest photocatalysts. Despite these successes, and taking into account the originality of the photocatalysts investigated, the research should also be viewed as a platform to be built upon, and therefore there are many possible follow up investigations that could be pursued in order to meet the 10 % solar to H_2 conversion efficiency using a robust system.

Chapter 3 detailed a facile method to control the exposing facet of Ag_3PO_4 and the influence on the activity. For any industrial application, Ag^+ cannot be used as an electron scavenger as metallic silver eventually poisons the surface, it is used here as a method to demonstrate a proof-of-concept in terms of potential maximum efficiency. Therefore, a sizable project would be to attempt to grow Ag_3PO_4 tetrahedrons onto a thin film, for use in a PEC cell. The contact between conducting substrate and photocatalyst would have to be exceptionally good to shuttle electrons around a circuit, otherwise as shown by Yi et al., in the absence of a suitable voltage, Ag_3PO_4 undergoes photocorrosion. Similarly, photocorrosion occurs over Ag_3PO_4 in the absence of soluble electron scavengers. It might also be possible to cover Ag_3PO_4 in a stable protection layer which is transparent to visible light, in order to act as an electron scavenger and prevent photocorrosion. Possibilities for cutting edge protection layers include; iron ferrihydrite (very high electronegativity), TiO_2 (robust over-layer), ZnO:Al or silica [4–6].

In Chap. 4, it was shown that by controlling the protonation status of g-C_3N_4, it is possible to increase the quantum efficiency up to 26 % under visible light. Previous studies have shown surface area to have an impact of sorts on the HER, however in comparison to altering the proton content, the increase in IQY is minimal. Investigations by Transient Absorption Spectroscopy (TAS) have previously provided understanding as to rate limiting steps in other photocatalysts such as TiO_2 and Fe_2O [7–9]. This same technique could be applied to carbon nitride to examine the kinetics of photogenerated charge carriers as a function of incident wavelength, which would illustrate a way to further improve the efficiency for the photocatalyst, e.g. by tailoring porosity and surface acidity.

Z-Scheme water splitting systems were studied in Chap. 5, and g-C_3N_4 was shown to be a functional photocatalyst for both oxidation of Fe^{2+} or I^- [10]. Despite the high quantum efficiency demonstrated in Chap. 4, in sacrificial systems, the STH% of the best Z-scheme system was low, at 0.1 %. Therefore, a further study would be required to probe this novel system. One route would be to experiment with alternative redox mediators, yet this would prove difficult as there are only a handful of mediators which are suitable for Z-Scheme water splitting. As mentioned, it is unknown whether the absorption of iodide or ferrous ions is surface-selective. Hence, another path to investigate would be an ion study on g-C_3N_4 synthesised using various precursors, possibly monitoring ion absorption using UV-Vis spectroscopy. Alternatively, it would also be possible to replace WO_3 with a photocatalyst which was more active in terms of oxygen evolution from water. The downside to this approach though, is searching for a photocatalyst which not only has visible light absorption and demonstrates a large OER, but also has a band gap which straddles the reduction potential of IO_3^-, and the redox potential of water, is indeed very difficult, but not impossible. It would also be worthy exploring the incorporation of other redox mediators into the overall water splitting systems, such as $NaNO_3$ or chromate complexes.

References

1. Yi, Z., et al. (2010). An orthophosphate semiconductor with photooxidation properties under visible-light irradiation. *Nature Materials, 9*, 559–564.
2. Wang, X., et al. (2008). A metal-free polymeric photocatalyst for hydrogen production from water under visible light. *Nature Materials, 8*, 76–80.
3. Martin, D. J., et al. (2014). Highly efficient Photocatalytic H_2 evolution from water using visible light and structure-controlled graphitic carbon nitride. *Angewandte Chemie International Edition, 53*(35), 9240–9245. doi:10.1002/anie.201403375.
4. Qu, Y., & Duan, X. (2013). Progress, challenge and perspective of heterogeneous photocatalysts. *Chemical Society Reviews, 42*, 2568–2580.
5. Liu, G., et al. (2014). A tantalum nitride photoanode modified with a hole-storage layer for highly stable solar water splitting. *Angewandte Chemie International Edition,*. doi:10.1002/anie.201404697.
6. Awazu, K., et al. (2008). A plasmonic photocatalyst consisting of silver nanoparticles embedded in titanium dioxide. *Journal of the American Chemical Society, 130*, 1676–1680.
7. Pendlebury, S. R., et al. (2011). Dynamics of photogenerated holes in nanocrystalline α-Fe_2O_3 electrodes for water oxidation probed by transient absorption spectroscopy. *Chemical Communications 47*, 716–718. doi:10.1039/C0CC03627G.
8. Tang, J., Cowan, A. J., Durrant, J. R., & Klug, D. R. (2011). Mechanism of O_2 production from water splitting: nature of charge carriers in nitrogen doped nanocrystalline TiO_2 films and factors limiting O_2 production. *The Journal of Physical Chemistry C, 115*, 3143–3150.
9. Tang, J., Durrant, J. R., & Klug, D. R. (2008). Mechanism of photocatalytic water splitting in tio2. reaction of water with photoholes, importance of charge carrier dynamics, and evidence for four-hole chemistry. *Journal of the American Chemical Society, 130*, 13885–13891.
10. Martin, D. J., Reardon, P. J. T., Moniz, S. J. A., & Tang, J. (2014). Visible Light-Driven Pure Water Splitting by a Nature-Inspired Organic Semiconductor-Based System. *Journal of the American Chemical Society, 136*(36), 12568–12571. doi:10.1021/ja506386e.